HOW SPACE CHANGED YOUR LIFE: FROM NASA LABS TO YOUR LIVING ROOM

THE SURPRISING STORY OF SPACE TECHNOLOGY IN HEALTH & HOME

NICOLA BIGGS

TWENTY SQUARES PUBLISHING

CONTENTS

Foreword v
Introduction vii

1. Landmark Achievements and Global Inspiration 1
2. Bringing Space Down To Earth 13
3. Understanding The View from Above 21
4. Invisible Infrastructure 30
5. Space Tech in Your Daily Life 42
6. Health and Medicine Insights From Orbit 52
7. Sleep, Stress, and the Space Mind 66
8. From Space to Plate 71
9. Space Partnerships 84
10. How Space Ideas Reach You 99
11. Debris and Dust 110
12. Conclusion 123

Bibliography 129

FOREWORD

Before you go any further, try to think about what space has done for you. Can you think about anything today that you have used that began, one way or another, on a spacecraft or in a space lab?

If you had to answer straight away, you might say the weather forecast, or satnav, or maybe not much at all. Most of us still think of space as something impressive but remote, something that belongs to astronauts, scientists, rockets, and Moon landings rather than ordinary life.

What we do not usually picture is the kitchen, the car, the doctor's office, the phone in a pocket, or the timing systems behind a network.

Because by the end of an ordinary day, you may well have used a whole chain of systems, products, and ideas that would not exist in the same form, or in some cases would not exist at all, without technologies first developed, tested, or accelerated by the effort to operate beyond Earth.

That is the territory this book sets out to explore.

INTRODUCTION

The short answer to the Foreword's question is that it's more than you think, and in ways you probably wouldn't guess.

This book explores how space exploration, especially the long public programmes led by agencies such as NASA, ESA, JAXA, and ISRO, has changed life on Earth in ways most of us rarely stop to notice.

These agencies, often operating with mandates extending to public benefit and supported by public investment, have been crucial in generating tangible, and often unexpected benefits, that directly improve and enrich our lives.

While the landscape is fast-moving with increasing commercial participation, understanding the unique role of this work is key to appreciating the full story. Because sending people and machines into the hostile environment of space isn't really about planting flags or capturing stunning images of distant stars.

It's about confronting extreme challenges that demand completely new solutions. For example, how do you protect astronauts from radiation? How do you generate power efficiently millions of miles from Earth? How do you create systems to recycle air and water for long-term missions? Or how do you build robots that can function by themselves on another planet?

And those innovations made to solve these problems rarely stay in space. They spread out, finding applications that improve our health, safety, connectivity, and understanding of our own world.

They have forced us to push the boundaries of science and engineering across countless fields from materials science, computing, robotics, communications, medicine, to environmental monitoring, and so much more.

This happens in many ways, influenced by whether the driving force is long-term public research goals or more immediate commercial objectives. The benefits are delivered in a variety of ways too - through patent licensing, open data policies, and collaborative partnerships that we'll examine in detail.

These are some of the themes that run through the chapters ahead.

Direct Applications
Technologies built for specific space challenges find practical uses in common items, industries, and infrastructure – from the memory foam in your mattress to the sensors in your digital camera, and advanced medical imaging techniques.

Scientific Understanding
Exploration deepens our knowledge, not only of the universe, but also of ourselves and our planet. Studying astronauts in microgravity provides unique insights into human physiology, while observing Earth from orbit offers deep data for weather forecasting, climate monitoring, and resource management many of which are shared openly as a public good.

Economic Engine
Sustained investment in space acts as a powerful driver of economic growth, driving technological advancement, supporting high-tech

jobs, creating new industries, and generating significant economic returns through specific mechanisms like technology transfer programs, patent licensing designed for broad adoption, and open data policies that fuel further innovation.

Serendipitous Discoveries

Sometimes, the most impactful benefits are entirely unexpected - happy accidents where solutions developed for one space problem find surprising applications elsewhere, or where basic research, often funded without immediate commercial application in mind, leads to unforeseen breakthroughs.

Inspiration and Achievement

Landmark missions, like the Apollo Moon landings or the voyages of robotic explorers to Mars, serve as powerful symbols of human ingenuity and perseverance, inspiring generations to follow careers in science and technology often achieved by international collaboration.

Ultimately, the journey is intrinsically linked to progress around us. Exploring space is a vital investment in developing the knowledge and technologies needed for that exploration, but it also helps to create new businesses, new profit streams and new jobs in a town near you.

The story of space exploration is, at the end of the day, our story.

1

LANDMARK ACHIEVEMENTS AND GLOBAL INSPIRATION

The best place to begin is with those big moments that most of us will remember, will have heard about, or will have watched a movie about. For many, they formed the lasting impression of what space travel was - flying a machine to far-away places to stand on new planets with astronauts dressed in puffy white suits with big helmets and black shiny visors.

I wasn't around for that first Moon landing, or the first flight to space, but I know almost everything about them.

But these are the cornerstones of the extraordinary journey that took humanity from our first tentative steps into space to establishing a permanent human presence among the stars.

They are things we tend to think of first when we hear the word spacecraft, spaceship or astronaut (which really does mean star sailor), but we rarely think about those who made it all happen.

These scientists and astronauts captured the world's imagination, united humanity in wonder, and pushed at the edges of what we thought possible.

The Beep Heard Round the World

For millions of people alive in 1957, the evening of October 4th was the first time that people listened to the world beyond us, when they first heard about. or even listened, to a simple, persistent radio signal coming from above.

"Beep... beep... beep..."

This steady 'beep' came from Sputnik 1, a 184-pound (83.6 kg) aluminum sphere launched by the Soviet Union. As it circled the globe every 96 minutes, this first artificial satellite marked humanity's entrance into the space-age with nothing more than its distinctive radio transmissions.

Today, the sight of a satellite crossing the night sky barely merits a mention. But in 1957, the un-crewed Sputnik's launch was a technological earthquake that changed geopolitical perceptions overnight. For Americans especially, the Soviet achievement was a shock that prompted immediate soul-searching. How had the U.S. fallen behind?

The response was swift and far-reaching. Within months, President Eisenhower signed the National Aeronautics and Space Act and NASA began operations in 1958. Congress also passed the National Defense Education Act, pouring unprecedented federal funding into science, mathematics, and foreign language education at all levels. A single space achievement had triggered a complete revolution in American education policy, whose effects are still felt today in how we teach STEM subjects.

The United States responded with its first satellite, Explorer 1, launched on January 31, 1958. Its primary goal was simply to put a U.S. satellite into orbit during the intense early days of the Space Race with the Soviet Union.

Dr. James Van Allen had managed to include a simple Geiger counter, designed to measure cosmic rays, on board the small satellite. The instrument returned puzzling data, periodically showing zero counts which initially looked like instrument failure.

But Van Allen correctly worked out that the counter wasn't malfunctioning; it was being overwhelmed by intense radiation that caused the detector to "saturate."

This led to the discovery of the radiation belts that now bear his name - the first major scientific discovery of the space-age, made possible by an instrument included almost as an afterthought.

More recently, NASA's Juno spacecraft, designed to study Jupiter's gravity and magnetic fields, made a surprise discovery too, about the origin of the zodiacal light. This faint, ethereal glow sometimes visible in the night sky after sunset or before dawn is sometimes called “false dawn”. It’s caused by sunlight scattering off tiny dust particles in the inner solar system, known as interplanetary dust.

Using its navigational star tracker and magnetometer (instruments intended for completely different purposes), the Juno team was able to analyze impacts from dust particles hitting the spacecraft.

To their surprise, the data suggested that this dust likely originates from Mars, kicked up by Martian dust storms and spreading throughout the inner solar system. This finding was completely unrelated to Juno's primary mission objectives but provided valuable insights into how dust moves through our solar system.

It helped explain what causes the faint glow of zodiacal light which meant that, if scientists know where the dust comes from, they can better understand why that glow appears.

It demonstrated that planets are not completely separate worlds and that dust from Mars may be travelling much farther than we had expected, which means that material can move around the Solar System more than people once thought.

The finding is useful for space missions too, because tiny dust particles can affect spacecraft and scientific instruments.

All of this shows us that sometimes instruments designed for one purpose reveal something entirely unexpected.

The Compton Gamma Ray Observatory is another example. Launched in 1991 to study distant high-energy phenomena like black holes and supernovae, it unexpectedly detected brief, intense flashes of gamma radiation coming from Earth itself.

These terrestrial gamma-ray flashes, now known to be associated with lightning storms, opened up an entirely new field of study in atmospheric physics that no one had anticipated when the mission was planned.

Similarly, the Kepler space telescope, designed to detect planets around other stars by measuring tiny dips in starlight, has also revolutionized our understanding of the stars themselves through asteroseismology (the study of stellar oscillations). This was a known potential secondary science objective, but the richness of the data and the insights gained into stellar interiors exceeded all expectations.

These satellites, born of Cold War competition, proved the feasibility of operating technology in orbit and became the basis for the communications, navigation, weather, and Earth observation satellites that are now so integral to your daily life.

When you check the weather on your phone or navigate using GPS, you're benefiting from the path taken by these first orbital explorers.

One Orbit Changes Everything

Imagine the courage it took to be sealed inside a small metal capsule, strapped to a converted ballistic missile, a type of rocket with a notorious record of explosive failures, and blasted toward the edge of space, knowing you'll be the first human to experience orbital flight.

On April 12, 1961, Soviet cosmonaut Yuri Gagarin did exactly that, becoming the first human to orbit Earth aboard the Vostok 1 spacecraft. His 108-minute flight transformed our species from Earth-bound to spacefaring.

His reported words as the rocket lifted off," Poyekhali!" ("Let's go!"), became an iconic declaration of human determination to explore beyond our home planet.

And, less than a month later, on May 5, 1961, Alan Shepard became the first American in space, though his Freedom 7 flight as part of Project Mercury in a shorter, suborbital journey. John Glenn followed on February 20, 1962, becoming the first American to orbit Earth aboard Friendship 7.

These pioneering missions provided the first data on how the human body reacts to spaceflight, revealing changes in cardiovascular function, muscle response, and how our balance system adapts to weightlessness. These discoveries, published in scientific literature, formed the basis of space medicine and contributed to our broader understanding of human physiology under extreme conditions.

The engineering challenges these missions overcame - from developing reliable life support systems, spacecraft controls, and re-entry methods - built expertise that would prove essential for more ambitious journeys. Each mission pushed further than the last, moving spaceflight from science fiction fantasy to operational reality.

The Apollo Odyssey

It is unlikely that any human endeavor of the 20th century will capture the global imagination quite like the race to the Moon.

On May 25, 1961, President John F. Kennedy made a commitment before Congress, declaring that the United States should "commit itself to achieving the goal, before this decade is out, of landing a man on the Moon and returning him safely to Earth."

At the time, the total American experience in space amounted to just

15 minutes of Alan Shepard's suborbital flight. The goal seemed almost impossibly ambitious.

To fulfil Kennedy's vision, NASA launched the Apollo program, an undertaking of unprecedented scale and complexity. It required mobilizing vast resources and pushing the capability of technology in countless areas: rocketry (resulting in the massive Saturn V, still the most powerful rocket ever successfully flown), navigation, computing, materials science, life support, and operations management.

On July 20, 1969, people around the world gathered to witness an achievement previously confined to the realm of science fiction. As Apollo 11 commander Neil Armstrong took his historic "one small step" onto the lunar surface, an estimated 600 million people (one-fifth of the world's population at the time) watched it on television in their living rooms, or crowded into a neighbour's house, a bar, a pub, a school hall, or a shop window to see it.

Just imagine how that moment might have affected a child watching it unfold, inspiring dreams that would change their future. Or how it demonstrated to people everywhere that seemingly impossible challenges could be overcome through human ingenuity and determination.

Apollo delivered significant scientific returns, including hundreds of kilograms of lunar samples that continue to inform our understanding of the Moon's origin and history. These materials, preserved and studied for over five decades, are still providing new insights as analytical techniques improve.

And the Apollo Guidance Computer, which took astronauts to the Moon and back, had roughly 100,000 times less processing power than the smartphone in your pocket. Yet it performed its critical functions with such reliability that it helped prove key principles of real-time, fault-tolerant computing. Those ideas later informed the systems which would eventually become the digital fly-by-wire technology that would revolutionize aircraft control systems.

The technological advancements spurred by Apollo had other lasting impacts far beyond NASA. For example, after the tragic Apollo 1 fire, NASA invested heavily in developing fire-resistant materials for spacesuits and spacecraft interiors. One such material, PBI (Polybenzimidazole), known for its exceptional thermal stability and flame resistance, was later commercialized and is now widely used in protective gear for firefighters, industrial workers, and military personnel.

When you see a firefighter rushing into a burning building, protected by their specially designed gear, you're witnessing one of Apollo's many legacies, and just one of the many technologies developed for space that now save lives on Earth.

Robot Explorers

While industrial robots and automata had existed for decades before the space-age, space exploration pushed robotics into a new category altogether - machines that had to operate themselves, far from help, in environments no human could survive. These sophisticated explorers have ventured to places we cannot yet reach, returning breathtaking images and invaluable scientific data that have transformed our understanding of our solar system.

In the 1960s and 70s, NASA's Mariner program provided the first close-up views of Venus, Mars, and Mercury, revealing landscapes both eerily familiar and utterly alien. The Viking landers touched down on Mars in 1976, conducting the first experiments searching for signs of life on another world.

The most ambitious planetary missions were the twin Voyager spacecraft, launched in 1977. Taking advantage of a rare planetary alignment that occurs only once every 176 years, these robotic explorers conducted what scientists called the "Grand Tour" of the outer planets - Jupiter, Saturn, Uranus, and Neptune - before continuing their journey into interstellar space, where they still operate today, almost 50 years later. They were clearly built to last.

The discoveries really were revolutionary. Voyager found active volcanoes on Jupiter's Moon Io, the most volcanically active world in the solar system, and the first volcanic activity observed beyond Earth.

It revealed the delicate structure of Saturn's rings and the strange tilted magnetic field of Uranus. It discovered geysers erupting from Neptune's Moon Triton. Each finding, published openly and shared globally, rewrote our understanding of these distant worlds.

More recently, missions like NASA's Cassini orbiter spent over a decade studying Saturn and its Moons, revealing the possible habitability of Enceladus with its subsurface ocean. The European Space Agency's Rosetta mission achieved the feat of orbiting and landing on a comet, providing unprecedented insights into these primordial bodies. Japan's Hayabusa missions successfully returned samples from asteroids in 2010, helping us understand the building blocks of the solar system.

NASA's work on autonomous rover navigation - machines that had to find their own way across Martian terrain without real-time human control - has helped design how self-driving cars make decisions in uncertain conditions.

And the haptic feedback systems developed so astronauts could 'feel' what a robot arm was gripping have helped modern surgical robotics, which we'll come back to in the chapters on medicine.

Living in Space

Today, astronauts routinely spend six months or more orbiting Earth, and the record (at time of writing) for a single spaceflight stands at 437 days, set by Russian cosmonaut Valeri Polyakov aboard the Mir space station.

The dream of a permanent human presence in space began to take shape with early space stations like the Soviet Union's Salyut series and NASA's Skylab in the 1970s. These were followed by the Soviet/Russian Mir station, which hosted crews continuously for nearly a decade.

Today, the International Space Station (ISS) stands as the pinnacle of this effort and a testament to global cooperation. A joint project involving NASA, Roscosmos (Russia), ESA (Europe), JAXA (Japan), and CSA (Canada), the ISS has been continuously inhabited since November 2000 with over two decades of unbroken human presence in orbit.

Next time you look up at the night sky, remember that right now, at this very moment, there are people living and working in space, circling our planet every 90 minutes at a speed of 17,500 miles per hour.

The ISS

The ISS serves as a unique microgravity laboratory where astronauts conduct research that would be impossible on Earth. Studies in biology, human physiology, materials science, Earth observation, and astronomy performed on the station have given us insights impossible to gain on our gravity-bound planet.

We cover many of these in the following pages, and just one example now is the advanced water purification systems developed for the ISS which are capable of recycling urine and wastewater into usable water.

Similarly, research conducted on the ISS has helped us understand much more about materials science, combustion, and human health with direct terrestrial applications. Studies on bone loss in microgravity, for example, have contributed to better understanding of osteoporosis.

The operational experience and technological advancements we have gained from building and maintaining this complex orbital outpost has provided invaluable lessons for future space travel while contributing directly to technologies applicable in your daily life.

The Expanding Space Community

If you're like most people, the United States and Russia (formerly the Soviet Union) are probably top your list when you think of countries that are involved in space. But the exclusive club of nations capable of putting satellites, probes, and humans into space has expanded dramatically in recent decades.

China has developed its own human spaceflight program (Shenzhou), launching lunar missions (Chang'e), building its own space station (Tiangong), and landing rovers on Mars (Zhurong). These achievements have established China as the third nation capable of independently launching humans into orbit.

India has emerged as a major space power, successfully launching lunar (Chandrayaan) and Mars (Mangalyaan) orbiters, developing its own satellite navigation system (NavIC), and achieving the first soft landing near the Moon's south pole with Chandrayaan-3 in 2023. What makes India's achievements particularly interesting is their cost-effectiveness; the Mangalyaan Mars orbiter, for instance, reached the Red Planet at a remarkably low cost compared to similar missions - with a reported cost of approximately $74 million.

The European Space Agency (ESA), representing 22 member states, has also established itself as a leader in Earth observation, science missions, and launch capability with its Ariane rocket family. Japan's space agency, JAXA, has pioneered innovative technologies like solar sails and asteroid sample returns.

Even smaller nations and private companies now have access to space. The United Arab Emirates successfully sent the Hope probe to Mars. South Korea, Brazil, and Israel have developed indigenous satellite launch capabilities. New Zealand became an unexpected space player with Rocket Lab, a private company offering dedicated small satellite launches. Even Scotland is due to have its own rocket launch site.

This expanding community of space participants brings a range of perspectives, innovative approaches, and new capabilities to humanity's collective space journey. It demonstrates that space exploration is truly becoming a global enterprise.

Inspiration as the Ultimate Spinoff

These landmark achievements, from the first satellite beeps to footprints on the Moon, robotic explorers venturing to distant worlds, and the ongoing international collaboration aboard the ISS, are powerful narratives of human curiosity, ingenuity, and perseverance.

They inspire students to pursue careers in science, technology, engineering, and mathematics. How many engineers, scientists, and entrepreneurs chose their paths after watching a space shuttle launch, seeing the Mars rovers in action, or learning about the cosmic discoveries of the Hubble Space Telescope?

Space achievements bring a sense of global unity when the world watches a shared human experience unfold. Remember the international excitement surrounding the Philae lander touching down on a comet? Or the global audience that watched NASA's Perseverance rover land on Mars in 2021? These moments cross national borders and remind us of our shared humanity and capacity for wonder.

While difficult to quantify, this inspirational value is a significant and lasting benefit of space exploration. The focused effort required to achieve these milestones consistently pushes the limits of science, creating a wealth of knowledge and capabilities, both planned and unexpected, that ultimately benefit society.

The achievements are the visible part. What came back with them - rocks, dust, fragments of asteroids, samples of comets - is where the next chapter begins.

2

BRINGING SPACE DOWN TO EARTH

In the summer of 2023, a capsule the size of a car tyre parachuted into the Utah desert. It had travelled nearly four billion kilometres to get there. Inside, sealed against contamination by Earth's air and bacteria, were 250 grams of material from an asteroid called Bennu. It was the largest sample of space rock ever returned to Earth by a NASA mission. Scientists had been working toward that moment for seven years.

It took time, but analysis of the sample found amino acids, water-bearing minerals, and uracil - one of the chemical components of RNA (the molecules that convert genetic information into the living components of your body). These molecules were preserved in rock that had changed little since the solar system formed 4.5 billion years ago.

That is what this chapter is about. Space research reaches us in many forms, through satellites, through materials, through data gathered from orbit. But the most direct route is the oldest one - you go there, you collect something, and you bring it back. What those returned samples have told us about our Moon, about the asteroids that cross

Earth's path, about where the ingredients for life may have come from, is the subject of the pages that follow.

Lessons from the Moon

The most iconic sample return effort remains NASA's Apollo program. Between 1969 and 1972, six missions landed astronauts on the Moon, and they returned with a collection of 842 pounds of lunar rocks, core samples, pebbles, dust, and soil.

This collection, gathered from various landing sites, overturned what we thought we knew about our nearest celestial neighbor and the early solar system, with direct implications for understanding Earth.

Before Apollo, theories about the Moon's origin abounded. Was it a captured asteroid? Did it form at the same time as Earth?

When researchers examined the isotopic composition of oxygen and other elements in Moon rocks, they discovered that these isotopic fingerprints were similar to those found in Earth's mantle.

By measuring the decay of radioactive isotopes within the rocks, a technique akin to a cosmic clock, they determined the Moon formed very early in the solar system's history, shortly after Earth itself. This discovery strongly supported what we call the Giant Impact Hypothesis - which said that the Moon was formed from debris ejected when a Mars-sized object collided with the young Earth about 4.5 billion years ago.

This happened in our solar system's violent youth, and not only did it gave us our Moon, but it may have tilted Earth's axis, creating our seasons - which has led to our own planet's unique development.

Analysis also revealed two distinct geological areas on the lunar surface. The dark, smooth 'maria', vast plains of ancient volcanic basalt flows, and brighter, heavily cratered highlands which are composed of older rocks like anorthosite - a light-colored rock made mostly of feldspar minerals.

Studying the ages of these different impact craters provided a timeline for the intense bombardment of the early inner solar system.

Since Earth would have experienced an even greater bombardment (being a larger target), understanding the Moon's impact history provides our best window into the conditions on early Earth, a period from which terrestrial rocks have largely been erased by geological activity.

This knowledge informs models of Earth's early atmosphere, ocean formation, and the conditions under which life arose.

Within the rocks, scientists also identified Helium-3, an isotope rare on Earth but potentially abundant on the Moon, which remains a long-term consideration for future fusion energy concepts.

The most unexpected thing the samples revealed, decades after their collection, was water.

While initially thought to be bone-dry, more sensitive analysis of Apollo samples, combined with remote sensing, revealed traces of water ice trapped in permanently shadowed craters.

Confirming the presence (and accessibility) of water is a key goal for sustainable lunar exploration, as it could provide drinking water, breathable air (by splitting H_2O), and rocket propellant - but the engineering challenges are still significant.

Decades later, we're still learning from the Apollo samples and thanks to the increasing sophistication of lab tools they are revealing details scientists could only dream of in the 1970s. It's like reopening a time capsule with better eyes every few years.

And lets not forget that studying the physical and chemical properties of lunar soil (regolith) with its abrasiveness, composition, and geotechnical properties, is essential for designing long-lasting rovers, habitats, and spacesuits. For example, understanding how abrasive lunar dust affects seals and mechanisms is vital for long-term hardware reliability.

Insights from Meteorites

While missions like Apollo provide targeted samples, nature delivers its own extraterrestrial materials to Earth in the form of meteorites.

Although a shooting star is a small piece of cosmic debris burning up in our atmosphere, some larger fragments make it all the way to the ground, becoming natural sample return missions without the billion-dollar price tag.

These meteorites are essentially "free" samples, though finding and analyzing them presents its own challenges. Most go unnoticed, landing in remote areas or the ocean. Those that are recovered provide a window into the composition of their parent bodies whether they be asteroids, space junk or pieces of Mars and the Moon.

Meteorites are broadly classified based on their make-up and they tend to be stony, iron, or stony-iron and this reflects the nature of their parent bodies. The most exciting discoveries, with direct relevance to the origin of life on Earth, have come from carbonaceous chondrites.

Analysis of these meteorites has found the presence of a wide range of organic molecules, including amino acids (the building blocks of proteins) and nucleobases (components of DNA and RNA).

Finding these essential ingredients for life within meteorites lends support to the theory that asteroids and comets may have delivered water and the chemical precursors for life to the early Earth, addressing a fundamental question about our own existence.

But meteorites also pose challenges. They are altered by their passage through Earth's atmosphere, and once they land, they are quickly contaminated by terrestrial microbes and weathering. Determining their precise origin is often difficult.

This underscores the value of pristine samples returned directly from known bodies by spacecraft, allowing for more definitive conclusions that might be relevant to planetary science and astrobiology today.

For this reason, in recent decades, space agencies have undertaken missions to collect samples directly from small bodies and return them uncontaminated to Earth. As well as providing details about the solar system's earliest ingredients they are also directly relevant to current challenges like planetary defense and future opportunities like resource use.

NASA's Stardust mission, launched in 1999, flew through the coma (the dusty halo) of Comet Wild 2 in 2004. Using a collector filled with aerogel, it captured thousands of tiny dust particles shed by the comet. As well as primitive materials the samples contained minerals formed at high temperatures, suggesting significant mixing occurred in the early solar nebula, pushing material from near the young Sun outward.

Crucially, Stardust samples provided the first definitive detection of the amino acid glycine in a comet, bolstering the idea that comets could have seeded early Earth with life's building blocks.

Japan's Hayabusa mission reached asteroid Itokawa in 2005 and became the first mission to successfully land on, take off from, and later return samples from an asteroid. The sample capsule came back to Earth in June 2010. Although technical problems meant the return was tiny, scientists were able to confirm that it contained microscopic dust grains from Itokawa. Later analysis showed that Itokawa is a "rubble-pile" asteroid, meaning it is made of loosely bound rocky material rather than being one solid monolithic body.

Its successor, Hayabusa2, went even further, reaching the carbon-rich asteroid Ryugu in 2018. After mapping the surface in detail, it performed not just one, but multiple sampling operations, including one that collected material from beneath the surface by creating a small crater with an impactor. The samples returned to Earth in 2020.

Analysis of the Ryugu samples revealed a treasure trove of information. The asteroid is rich in water-bearing minerals and contains a number of organic molecules, including multiple types of amino acids and even uracil, one of the nucleobases in RNA. These findings strongly suggest that C-type asteroids like Ryugu could have been a significant source of water and organics for early Earth.

NASA followed with its OSIRIS-REx mission, which collected a substantial sample, estimated at around 250 grams, from another carbonaceous asteroid, Bennu, in 2020, returning it successfully in 2023. Bennu is classified as a potentially hazardous asteroid, meaning its orbit brings it relatively close to Earth.

Studying the returned sample is therefore crucial for planetary defense because it will mean that scientists can better predict how such an asteroid might react to an impact or other deflection techniques, should one ever be necessary.

Beyond the origins-of-life implications, understanding the composition of asteroids like Bennu and Ryugu is also important for space resource utilization (SRU).

These bodies contain water and potentially valuable metals and the samples provide detailed "ground truth" about the concentration and accessibility of these resources. This all helps to work out the feasibility and design of future mining or extraction missions.

These missions, targeting primitive bodies that have changed little since the solar system's formation, provide invaluable clues about the materials that formed the planets, and the role of water and organics in the early solar system.

That is why Mars sample return has long been seen as one of the great prizes of planetary science. A carefully selected sample from Mars could answer questions no rover can fully answer on its own, including the planet's past habitability and whether any traces of ancient life might still be preserved in its rocks.

It means that, instead of inferring everything from a distance, scientists can test real material in laboratories on Earth using tools far more powerful than anything a spacecraft can carry. That gives us a better understanding of how planets formed, how the ingredients of life may have spread through the early solar system, and how potentially hazardous asteroids are actually put together. It also pushes the development of analytical techniques that later find uses in medicine, forensics, and materials science.

Tiny Treasures

Because samples are scientifically priceless, scientists must develop analytical techniques that are extraordinarily sensitive, capable of analyzing minuscule amounts of material, and often non-destructive, so that they preserve the sample for future studies with even more advanced tools.

Techniques like high-resolution electron microscopy, which allow scientists to image sample structures down to the atomic scale with advanced mass spectrometry, helps to date and trace their origins.

Meeting this challenge has driven advancements that directly benefit numerous fields for us right now.

In medicine and forensics, improvements in mass spectrometry contribute directly to medical diagnostics. They have led to more sensitive disease marker detection, newborn screening, drug analysis and forensic science (like trace evidence analysis that can solve crimes).

For materials science and industry, advanced microscopy techniques honed for examining the nanostructure of lunar rocks or asteroid fragments are used to develop new materials, analyze failures in industrial components, and improve quality control in semiconductor manufacturing.

Environmental monitoring also benefits. The techniques developed to detect minute traces of specific elements or isotopes in extraterrestrial samples can be adapted for ultra-sensitive tracking of pollutants or understanding geochemical cycles on Earth.

Meanwhile, the need to prevent terrestrial contamination has helped advancements in cleanroom technologies and sample handling protocols in pharmaceutical production, biotechnology labs, and microelectronics.

The amino acids in the Ryugu sample, the water trapped in the Apollo regolith, the glycine in the comet dust were not what the missions went looking for. They were answers to questions nobody had thought to ask yet, and found in material collected for entirely different reasons. Space science has a long history of that kind of discovery, and it runs through much of what this book covers.

The samples will keep giving as new techniques are developed and we will find things in them that current methods can't see. When future missions return material from Mars, or from asteroids not yet named, those samples will join the same archive and provide physical evidence waiting for our questions.

3

UNDERSTANDING THE VIEW FROM ABOVE

While much of space exploration involves looking toward distant stars and planets, some of its biggest contributions come from turning our gaze back toward home.

Most of us will check the weather forecast on our phone before heading out for the day. And some of us may even track a hurricane's path as it approaches the coast, or will have admired the images showing the retreat of Arctic sea ice.

This is all thanks to hundreds of satellites circling our planet, their electronic eyes constantly watching and measuring our world from above. They orbit above Earth, watching weather systems move across continents and tracking changes in forests, oceans, and the energy that drives our climate.

Your Weather Forecast

Think back to how people predicted weather before the space-age. Without satellites, meteorologists relied on scattered ground observations, weather balloons, and a healthy dose of educated guesswork. Forecasting beyond a day or two? Nearly impossible. A storm forming over the open ocean might make landfall with little warning, giving coastal communities very little time to prepare.

Everything changed in 1960 when NASA launched TIROS (Television Infrared Observation Satellite), the first weather satellite. Suddenly, meteorologists could see entire weather systems developing and moving across the planet.

Today, a sophisticated network of satellites operated by agencies around the world keeps constant watch on our atmosphere. NOAA's GOES satellites hover in geostationary orbit 22,300 miles above specific points on Earth, providing those familiar images you see on weather broadcasts. The European Organisation for the Exploitation of Meteorological Satellites (EUMETSAT) operates the Meteosat series with similar capabilities over Europe and Africa. Japan's Himawari satellites watch the western Pacific, while India's INSAT series monitors the Indian Ocean region.

As well as taking pictures, these satellites carry instruments that measure temperature at different atmospheric levels, track water vapor, detect cloud heights, observe wind patterns, and much more - all feeding into complex computer models that simulate how the atmosphere will behave in the coming hours and days.

The five-day forecast today is as reliable as the three-day forecast was just a decade ago. This is the direct result of better satellite data being fed into increasingly sophisticated weather models.

When a hurricane forms in the Atlantic, satellites track its every move. They measure the storm's structure, estimate its intensity from cloud temperatures, and observe the surrounding atmospheric conditions that will steer its path. According to NOAA, the average error in a three-day hurricane track forecast has decreased by roughly 65% since 1990, a period marked by significant advances in satellite technology. Those extra days of accurate warning give communities crucial time to prepare or evacuate, saving countless lives and preventing billions in property damage.

Even your weather app relies heavily on satellite data about cloud cover, temperature, humidity, and wind patterns. Airlines use this information to plan fuel-efficient routes around turbulence. Farmers decide when to plant or harvest based on expected rainfall. Energy companies predict heating or cooling demand. Shipping companies navigate safely.

And this data is typically shared freely among nations through agreements coordinated by the World Meteorological Organization. Your weather forecast represents international cooperation at its finest - a global public service built on publicly funded infrastructure delivering tangible benefits every single day.

Taking Earth's Temperature

Climate change presents one of humanity's greatest challenges. But how do we actually know what's happening to our planet on a global scale? You've probably seen graphs showing rising global temperatures or shrinking polar ice caps, but where does this data come from?

The answer, increasingly, is space. Understanding climate requires consistent, precise measurements spanning the entire globe over many decades, a task perfectly suited for satellites. Space agencies worldwide operate specialized missions, often working together, to monitor key indicators of Earth's changing climate with unprecedented accuracy.

NASA's Earth Observing System includes satellites like Terra, Aqua, and Aura, carrying suites of instruments that measure everything from atmospheric composition to land surface temperature and vegetation health. Ever heard of the "ozone hole"? Satellites discovered it and continue to monitor its fluctuations.

If you have wondered how fast sea levels are rising then the joint U.S./European Jason series of satellites uses radar altimetry to measure global sea level - down to millimeters. These measurements have revealed that sea level rise is increasing, a critical finding for coastal communities worldwide.

Europe's ambitious Copernicus program, managed by ESA and the European Union, provides comprehensive environmental monitoring through its Sentinel satellites. Sentinel-1 uses radar to see through clouds and darkness, crucial for monitoring sea ice year-round. Sentinel-2 captures detailed optical imagery for tracking land use changes and forest health. Sentinel-3 monitors oceans, measuring temperature, color, and wave heights. Sentinel-5P specifically tracks air pollution and greenhouse gases, and it can even pinpoint specific factories or power plants emitting excessive pollutants.

And if you ever wondered how scientists track greenhouse gases like carbon dioxide and methane then Japan's GOSAT (Greenhouse gases Observing Satellite) and its successor GOSAT-2 precisely measure these gases globally, helping identify both sources (like industrial areas) and sinks (like forests absorbing carbon dioxide).

What makes this data particularly valuable is that much of it is freely available to anyone through open data policies championed by these government-funded agencies. NASA's Earth Observing System Data and Information System (EOSDIS), the Copernicus Open Access Hub, and JAXA's G-Portal provide treasure troves of environmental data accessible to researchers, policymakers, businesses, and curious citizens alike.

This wealth of satellite data forms the bedrock of modern climate science. It allows researchers to track global trends, validate and improve climate models, understand complex interactions within the Earth system (like how clouds affect warming), and provide policymakers with objective, evidence-based information needed to address climate challenges.

Without this continuous, global, space-based monitoring, our understanding of our planet's climate system would be vastly poorer. And maintaining these crucial measurements over decades—the time scale needed to distinguish climate trends from natural variations—depends directly on sustained public funding commitments.

Satellites to the Rescue

Imagine you're an emergency manager responding to a major flood. Roads are underwater. Communications are down. You need to know: Which areas are hardest hit? Where are people stranded? What evacuation routes remain open? How is the situation evolving?

Satellites have changed disaster response by providing this critical information when it's most needed. When earthquakes level buildings, floods inundate communities, wildfires rage through forests, or volcanoes erupt, Earth observation satellites become lifesaving eyes in the sky.

After the devastating 2010 earthquake in Haiti, satellite imagery helped map the extent of damage across Port-au-Prince when ground assessment was nearly impossible. During major hurricanes, radar satellites that can see through clouds map flooding in real-time, showing emergency managers which neighborhoods need immediate evacuation. When wildfires spread across California or Australia, thermal sensors on satellites detect hot spots through smoke, helping fire crews track the advancing front and allocate resources effectively.

What makes this capability even more impressive is how the international community comes together to provide this critical information. The International Charter "Space and Major Disasters," initiated by ESA and the French space agency CNES and now involving numerous space agencies worldwide, ensures that satellite imagery is rapidly acquired and made freely available to emergency responders during crises.

Each participating agency contributes its unique capabilities. Radar satellites like ESA's Sentinel-1 or India's RISAT can see through clouds and smoke, crucial during floods or volcanic eruptions when visual satellites are blinded. High-resolution optical satellites like India's Cartosat series or commercial systems can provide detailed damage assessments after earthquakes or tsunamis.

Think about the difference this makes: In 1970, when a cyclone struck what is now Bangladesh, the world barely knew what was happening until after the storm passed, contributing to a tragic death toll estimated between 300,000 and 500,000 people. Today, satellites track every tropical cyclone from formation to landfall, providing days of advance warning and helping emergency managers make life-saving decisions based on objective, up-to-date information.

From Farm to Forest

The benefits of the view from above extend far beyond emergency situations. Farmers, foresters, water managers, and urban planners increasingly rely on satellite data to make better decisions about our natural resources.

Many modern farmers use what's called "precision agriculture" - using satellite imagery combined with GPS to manage their fields with unprecedented accuracy. By analyzing vegetation indices derived from satellites like Landsat or Sentinel-2, they can identify variations in crop health across their fields, applying water, fertilizer, and pesticides only where needed. This not only reduces costs but also minimizes environmental impact while increasing yields.

Foresters use similar technology to monitor vast wooded areas impossible to survey comprehensively from the ground. They track deforestation and illegal logging (particularly important in remote rainforests), monitor forest health, assess fire risk based on vegetation dryness, and even estimate carbon storage - critical information for climate management strategies.

Water resource managers, especially in drought-prone regions, rely heavily on satellite measurements. They use satellite data to monitor snowpack (which provides much of the western United States' water supply), soil moisture content, reservoir levels, and evaporation rates. This information helps them predict water availability, allocate supplies efficiently between competing needs like agriculture and cities, and implement conservation measures before shortages become severe.

Urban planners use high-resolution imagery to map city growth, identify areas at risk from flooding or landslides, monitor urban heat islands (where cities trap heat, exacerbating hot weather), and track infrastructure development. For rapidly growing cities in developing nations, where official maps often lag far behind actual development, satellite imagery provides the most accurate picture of urban expansion.

India's space agency, ISRO, demonstrates how these capabilities benefit developing economies. Its Cartosat series provides high-resolution imagery vital for mapping infrastructure, planning development, and managing natural resources within India. Its RISAT (Radar Imaging Satellite) series provides all-weather imaging capabilities crucial for monitoring agriculture during monsoon season, when clouds would obscure optical satellites.

The data enabling these applications often comes from long-running government satellite programs. The Landsat program, the joint NASA/USGS effort, has provided continuous imagery of Earth for nearly 55 years, creating an unparalleled record of our changing planet. The European Copernicus program's Sentinel satellites continue this tradition with enhanced capabilities, higher resolution, and more frequent coverage.

Open Data and Public Investment

If you ever used Google Earth to zoom in on your neighborhood or a vacation destination then you have probably used global imagery. These tools largely exist because of a policy decision of many government-funded space agencies who decided to make their Earth observation data freely available to anyone.

This approach, of treating Earth observation data as a public good, like weather information, multiplies the value of the initial public investment in these satellite systems. Researchers worldwide can access these vast archives to study everything from urban development patterns to ecosystem changes, climate trends to agricultural productivity. Businesses build innovative applications and services using this foundational data—creating jobs and economic activity far exceeding the original investment.

When the U.S. government made the entire Landsat archive freely available in 2008, usage exploded from a few thousand images per year to millions. Entrepreneurs created new services using this data, from agricultural monitoring platforms to forest management tools. Scientists conducted studies that would have been financially impossible if they had needed to purchase each image. Government agencies at all levels gained access to consistent, reliable environmental information to inform policy decisions. We talk more about Landsat later in the book.

The European Copernicus program took this even further, making all data from its Sentinel satellites freely accessible through its Open Access Hub. This policy has enabled thousands of businesses, researchers, and public agencies to utilize this valuable information for applications ranging from maritime safety to air quality monitoring, crop forecasting to infrastructure planning.

The data enabling these applications comes from a long-running government programme, and what happened when that data was made freely available to the world is covered in Chapter 10.

While private companies also operate Earth observation satellites, often providing very high-resolution imagery or specialized capabilities, they typically sell their data or derived products. The foundational, continuous, global monitoring provided by publicly funded missions with open data policies serves a distinct and critical role, ensuring broad, equitable access to essential environmental information.

Seeing Our Connected Planet

The next time you check a weather forecast, see a satellite image of a hurricane on the news, or use an app that incorporates environmental data, remember that you're benefiting from a global infrastructure that gives us an unprecedented view of our planet.

From space, we can see Earth as the interconnected system it truly is - where weather systems don't stop at borders, where ocean currents connect distant shores, where forest fires in one region affect air quality thousands of miles away. This perspective has fundamentally changed how we understand our world and how we manage our relationship with the environment that sustains us.

The collaborative efforts of international space agencies from NASA, ESA, JAXA, ISRO, NOAA, EUMETSAT, and many others, and their commitment to sharing data, underpinned by public investment, ensure that the insights gained from this orbital vantage point benefit all of humanity.

They help us understand, manage, and protect our home planet in ways impossible just a few decades ago. Most of that infrastructure is invisible to the people who depend on it every day.

4

INVISIBLE INFRASTRUCTURE

Most of what space has given us, you can't see.

The weather forecast on your phone is coming from a satellite. The directions your car is giving you are coming from another one. The payment you made at the coffee shop was time-stamped by a signal that bounced off an atomic clock in medium Earth orbit. The solar panels on your neighbour's roof exist in part because a very small satellite in 1958 proved that photovoltaic cells could power something in space.

The processor in your phone traces part of its ancestry to a decision made in 1962 about how to build a computer small and reliable enough to guide astronauts to the Moon.

Launching something above the atmosphere, keeping it alive in vacuum and radiation, tracking it precisely once it was up there, receiving and decoding its signals on the ground - none of that existed in 1957 when the space-age had barely begun.

Sputnik 1, the Soviet satellite launched that year, shocked the United States into action, and NASA was created the following year as part of that response.

All of it had to be invented. And because the problems were expensive, risky, and far beyond what any single company would undertake on its own, the learning happened inside national space agencies paid for by the public. In the United States, that mostly meant NASA, working alongside the Department of Defense, the Navy, and a network of university laboratories and industrial contractors.

That learning is the foundation that everything else sits on. Every satellite that has ever flown - military, commercial, scientific, the ones powering your phone's GPS, the ones carrying the TV broadcast, the ones watching the weather - flew on launch systems, operated in orbits, used tracking networks, and relied on timing and communications techniques that were worked out first by public space programmes. The commercial satellite industry exists because NASA and its counterparts around the world figured out some of the hard parts.

This chapter looks at four technologies where that is clearest. The positioning system your phone uses, the integrated circuits it is built from, the solar cells that power satellites and increasingly houses, and the communications satellites that carry a significant share of the world's information traffic.

GPS: Where You Are, To Within A Few Feet

In May 2000, the Clinton administration switched off a feature that changed daily life for most people in the developed world, and hardly anyone noticed.

Until that day, the Global Positioning System had been a US military asset that civilians could use only in a degraded form. The Department of Defense ran a feature called Selective Availability, which added random error to the civilian signal so that non-military receivers could only determine their location to within about 100 metres. Accurate enough to get you to the right town. Not accurate enough to navigate to a specific address.

That order turned Selective Availability off. Overnight, civilian GPS receivers became roughly ten times more accurate, locating themselves to within about 10-20 metres or better, and within a few years, the price of a GPS chip fell from around $3,000 to around $1.50. That combination of better accuracy, and cheaper hardware, is what put GPS into your phone, your car, and most of the infrastructure that modern life depends on.

The system itself had been under development since the early 1970s, consolidating earlier navigation satellite work by the US Navy (Transit, Timation) and the Air Force (Project 621-B) into a single programme called Navstar. The first Navstar satellite went up in 1978. The full constellation of twenty-four satellites became operational in 1995.

The engineering challenge was daunting: each satellite had to carry an atomic clock precise to a few billionths of a second, because GPS works by measuring tiny differences in how long radio signals take to travel from satellites to receivers. A clock error of one microsecond, multiplied by the speed of light, produces a position error of three hundred metres.

At the level of precision GPS needs to function, even Einstein's general relativity has to be taken into account. Clocks in orbit run about thirty-eight microseconds per day faster than clocks on the ground because of weaker gravity and orbital motion, and without correcting for this, position errors would accumulate at roughly ten kilometres per day. GPS receivers apply the correction automatically. It's one of the very few places in daily life where relativity is an engineering parameter rather than a page in a textbook.

The satellites themselves orbit at about 20,200 kilometres, much higher than the International Space Station. Your receiver - whether in your phone, your car, or your watch - listens for signals from at least four of them simultaneously. By measuring how long each signal takes to arrive, it can calculate its distance from each satellite, and from those four distances it can solve for its exact position and the current time. This technique, called trilateration, is the mathematical heart of the system.

GPS now does far more than help you find your way to a restaurant. Aviation depends on it for approaches to runways and shipping uses it to route container vessels through sea lanes. In farming, modern combine harvesters use GPS-guided steering to plant rows within a few centimetres of each other meaning that a substantial share of North American farmland is now cultivated using some form of GPS-guided equipment.

Then emergency services find callers more quickly because GPS locates them and financial markets use GPS timestamps to order transactions in a global system where microseconds can matter. Even power grids synchronise themselves using GPS time signals.

And three other systems now operate in parallel with the US one. Russia's GLONASS has been operational since the 1990s. Europe's Galileo reached full service in 2020. China's BeiDou achieved global coverage in 2020. Most modern smartphones listen to all of them at once, which is why your phone can find itself quickly even in a city centre where buildings block some satellites. The collective term for all of these is GNSS - Global Navigation Satellite Systems - and the existence of four overlapping constellations means the world now has genuine redundancy in one of its most critical infrastructure systems.

There's one feature of GPS you might not know and it's unusual in modern infrastructure. You don't pay for it. The US government chose, and has continued to choose, to provide the civilian signal free to the world. It's one of the largest and most valuable public goods ever created by any government.

The Apollo Computer

When NASA was planning the Apollo missions in the early 1960s, one problem stood out as almost unsolvable. The Apollo spacecraft would need to perform complex real-time calculations - orbital trajectories, engine burns, navigation - in an environment where a wrong answer could kill the crew. But computers in 1962 filled rooms. They weighed tons. They were nowhere close to fitting inside a spacecraft.

The engineer leading the effort at MIT's Instrumentation Laboratory, Eldon Hall, made a decision that many of his colleagues thought was reckless. He chose to build the Apollo Guidance Computer using an emerging and largely untested technology called the integrated circuit - a single piece of silicon containing multiple electronic components. Integrated circuits had been invented only a few years earlier, by Jack Kilby at Texas Instruments in 1958 and independently by Robert Noyce at Fairchild Semiconductor in 1959. By 1962 they were expensive, their reliability in complex applications was unproven, and nothing remotely as complex as a guidance computer had ever been built from them.

Hall went ahead anyway. The first version of the Apollo Guidance Computer used 4,100 integrated circuits from Fairchild, each containing a single three-input NOR gate. The later version used around 2,800 dual NOR gate chips, in the end supplied mostly by Philco-Ford, which had licensed the design. Around 800,000 integrated circuits in total were used across the Apollo programme. Every one was tested under rigorous conditions of temperature, vibration and contamination. If any chip from a batch failed, the whole batch was discarded. No Apollo Guidance Computer ever failed during a mission.

The significance of this wasn't only that it made Apollo possible. It was that in 1963, NASA's Apollo programme was consuming around 60 per cent of all integrated circuits being manufactured in the United States. The Department of Defense was buying most of the rest for the Minuteman II missile. Between them, these two programmes were essentially the entire market for integrated circuits at exactly the moment when the industry most needed a reliable buyer. The production runs, the manufacturing experience, the reliability testing, the price reductions that came with scale - all of this was happening because Apollo and Minuteman were paying for them.

By the end of the 1960s, integrated circuits had fallen in price enough to be used in commercial applications. The area of Santa Clara County where Fairchild and its competitors were based began to be called Silicon Valley. Fairchild itself spun off more than a dozen new companies, the most famous of which was Intel, founded in 1968. The industry that would produce every processor in every computer and phone you've ever owned had, by then, found its footing. Apollo was not the only reason. But it was a major one.

The Apollo Guidance Computer also forced the invention of modern software engineering. Margaret Hamilton, who led the software team at MIT, coined the term "software engineering" during the Apollo programme precisely because what her team was doing had no precedent.

They had to write code that would run in real time on an embedded computer with no room for bugs, handle multiple tasks simultaneously, and recover gracefully when things went wrong. During the Apollo 11 lunar descent, the computer started throwing overflow errors as it was overwhelmed with data from a radar switch left in the wrong position.

Because Hamilton's team had built in a priority interrupt system that allowed critical tasks to take precedence over less important ones, the computer simply restarted the most important processes and kept flying. The landing continued. That kind of fault-tolerant, priority-based real-time computing is now standard in everything from aircraft autopilots to medical devices to car engine management systems.

The computer in your pocket has maybe a hundred million times the processing power of the Apollo Guidance Computer. But the principles of how it organises its tasks, the reliability standards its chips are tested to, and the commercial ecosystem that manufactured it all trace back, in part, to decisions made about how to land twelve people on the Moon.

Solar Power, Where It Started

On 17 March 1958, the United States launched a small aluminium sphere called Vanguard 1 (and Soviet Premier Nikita Khrushchev dismissed it as "the grapefruit satellite" because it was so small). It weighed about 1.5 kilograms and carried two radio transmitters. One transmitter was powered by a conventional mercury battery while the other was powered by six small silicon solar cells mounted on its surface.

The battery-powered transmitter ran for about three months before the battery died. The solar-powered transmitter kept working for more than six years. Vanguard 1 proved that photovoltaic cells could produce useful electrical power in space over long periods - and that a satellite didn't have to die when its batteries did.

Silicon solar cells had been invented at Bell Laboratories only four years earlier, in 1954, by Calvin Fuller, Gerald Pearson and Daryl Chapin. The first Bell solar cell was about 6 per cent efficient at converting sunlight to electricity. It was a scientific curiosity. The cost per watt of electricity it produced was astronomical compared to burning coal. Nobody could imagine a commercial use on Earth.

Space changed that. In orbit, where there's no atmospheric dust, no night that lasts longer than 45 minutes, and where the alternative - batteries alone - means a satellite's useful life measured in weeks, solar cells were suddenly the only sensible choice. By the time Vanguard 1 was being designed, the Signal Corps had already concluded that photovoltaic cells were "most promising" for long-duration satellites. After Vanguard's success, solar power became standard for almost every satellite that has flown since.

This mattered beyond space in two ways. First, space was the only early market where cost per watt was irrelevant compared to reliability and weight. Through the 1960s, 1970s and 1980s, the space industry was the major customer driving improvements in solar cell efficiency and manufacturing methods. Efficiencies rose from around 10 per cent in the 1950s to around 30 per cent by the 2010s for the multi-junction cells used on modern spacecraft. Second, the manufacturing methods learned from making solar cells for space flowed back into terrestrial photovoltaic manufacturing. The silicon cells on rooftops today are a different product from those on spacecraft, but they owe part of their existence to decades of industry experience that began with satellites.

Solar power is now competitive with fossil fuels across much of the world. The International Energy Agency reports that solar is the cheapest source of new electricity generation in most countries. It's producing a growing share of the world's power, including on buildings that have nothing to do with space. The path from Vanguard 1 to a rooftop in Kent or California is long and has many contributors, but it is real.

Satellite Communications

On 10 July 1962, an American satellite called Telstar 1 was launched into an elliptical orbit around the Earth. It was a sphere about 90 centimetres across, covered in 3,600 solar cells. Built by Bell Laboratories, funded by AT&T, and launched by NASA, it was the first active communications satellite: a device in orbit that could receive radio signals from the ground, amplify them, and retransmit them back down.

Twelve days after its launch, Telstar 1 relayed the first live television signal across the Atlantic Ocean. On 23 July 1962, viewers in Europe and North America watched the same live broadcast simultaneously for the first time in history. Walter Cronkite appeared from New York. Images of the Statue of Liberty and the Eiffel Tower flashed across screens on both sides of the Atlantic. Before Telstar, television programmes produced in the United States had to be recorded on film or videotape and flown across the ocean by aeroplane, arriving days later.

Telstar itself was a brief experiment. Its electronics were damaged by radiation from an American high-altitude nuclear test, and it went out of service within a few months. But it had proved the concept.

NASA's own Syncom programme, which followed almost immediately, proved the next one. Syncom II, launched in 1963, showed that geosynchronous communications could work in practice.

Syncom III carried live television coverage of the 1964 Tokyo Olympics across the Pacific to North American viewers, the first time an intercontinental live broadcast had been possible in that direction. Geostationary communications satellites - satellites placed in orbits 35,786 kilometres above the equator, where they complete one orbit in exactly one day and therefore appear to hover over a fixed spot on Earth - followed within a few years. Intelsat I, launched in 1965, was the first commercial geostationary communications satellite. Intelsat IV, launched in 1971, could carry thousands of simultaneous phone calls and twelve colour television channels.

By the 1980s, satellite communications were carrying a large fraction of the world's international phone calls and nearly all live television broadcasts between continents. By the 2000s, they were routing television programming, radio broadcasts, corporate data networks, ship-to-shore communications, and the internet connections for aircraft, ships, and remote communities.

NASA itself built a parallel communications system for its own operations. The Tracking and Data Relay Satellite network, which has been in continuous service since the 1980s, is how the agency stays in contact with the Space Shuttle when it was flying, with the Hubble Space Telescope, and with the International Space Station. It isn't something most people will ever use directly, but it's an example of how satellite communications became an operational requirement for the space programme itself, not just something the programme helped create.

The newest generation of communications satellites doesn't sit in geostationary orbit. It sits much lower, in constellations of thousands of small satellites that work together. Starlink, operated by SpaceX, is the largest, with several thousand operational satellites providing internet service in more than a hundred countries. OneWeb, Amazon's Project Kuiper, and China's Guowang and Qianfan constellations are building similar systems. These low Earth orbit constellations can provide lower-latency internet service than geostationary satellites, because the signal has a shorter distance to travel. They've extended broadband internet access to rural communities, ships, aircraft, and aid organisations working in areas with no terrestrial infrastructure.

This expansion is the subject of a later chapter, because it raises real questions about debris, regulation, and who benefits from the new constellations. But the underlying infrastructure - the idea that radio signals can be relayed through space to connect any two points on Earth - traces back to the black-and-white transmission Telstar made in July 1962.

How and Why

These four technologies - GPS, integrated circuits, solar power, and satellite communications - don't have much in common on the surface. They solve completely different problems. They reach you through different industries. You interact with them at different times of the day.

But in each case, a technology that was either just emerging or had not yet found a commercial use was made practical by the demands of a space mission and the space programme was the early customer. It paid for the production runs, funded the reliability testing, drove the costs down, and proved that the technology could work at scale.

Space was a frontier that no one could reach without first learning how to operate there. The companies and agencies that eventually built commercial satellites, consumer GPS receivers, rooftop solar panels, and smartphone processors didn't step into a ready-made world.

They entered one where the launch vehicles, the tracking networks, the orbital mechanics, the radiation-hardened electronics, the precision timing and the manufacturing methods had all been developed first inside space programmes.

NASA was not the only institution involved. As we mentioned earlier, the Department of Defense, the Navy, the Air Force, and international partners all contributed, but NASA was central. And the knowledge it built was, by design, published, licensed, and shared.

That sharing is the mechanism by which space research turned into everyday life. NASA's procurement decisions told the semiconductor industry there was a buyer for what it was trying to build. NASA's open publication told the global research community what had worked and what hadn't. NASA's technology transfer programme, which we'll look at properly in a later chapter, moved specific inventions out of the laboratory and into the commercial world. Other agencies elsewhere did similar work. Together, they built the environment in which the commercial satellite economy, and much of modern electronics, could emerge.

This is different from the spinoffs covered in later chapters. Memory foam is a product that can be traced to a specific NASA project. What we've looked at here isn't a list of products but the foundational layer of modern infrastructure - the computing, the positioning, the communications, the power generation - and the role space programmes played in getting that infrastructure to the point where it could support everything else.

This is what makes space investment hard to evaluate. If you ask "what has the space programme given me?", the honest answer includes not just the items on your shelf but much of the system that allows a modern city to function. The electricity, the communications, the timing signals, the imagery, the logistics - all of it depends on infrastructure that space either created or helped create.

Most of the time, you don't see it. That's the nature of infrastructure. It only becomes visible when it fails.

5

SPACE TECH IN YOUR DAILY LIFE

Because of the number of pathways like patent licensing, open data policies, and industry partnerships (that we will explore later), some of the most extraordinary innovations of space exploration are now in your bedroom, kitchen, garage, and glasses case.

This chapter is about the tangible, practical, engineered tools and materials around you right now and each was born from a problem that only space could have posed.

The Mattress

In the 1960s, engineers at NASA's Ames Research Center faced a problem: how to protect astronauts and test pilots from the crushing G-forces of launch and landing.

Existing cushions compressed too quickly and offered uneven support. Charles Yost, a contract engineer working with NASA, developed a viscoelastic polyurethane foam that responded differently. Instead of springing back instantly, it deformed slowly and evenly under pressure, then gradually returned to its original shape.

This new "temper foam" absorbed shocks and distributed pressure more effectively than anything before it. In tests, it showed that it could soften to fit the body, reduce impact, and return to shape without damage. For pilots and astronauts, this was life-saving. But its use didn't stop there.

In the 1980s, NASA made the formulation available through its systematic technology transfer program (more in a later chapter). The medical community adopted it first, using it in wheelchairs and hospital beds to prevent pressure sores. Later, companies licensed the material for consumer products. By the 1990s, it was appearing in mattresses, pillows, athletic shoes, and protective sports gear.

Today, memory foam is everywhere, from pillows and mattresses to medical cushions. But the principle behind it of managing force through slow, even absorption was developed because engineers needed a material for astronauts that could absorb launch and landing forces.

Cordless Tools

Drilling a core sample on the Moon required a tool unlike anything then in common use. There were no power sockets on the lunar surface, and cords would tangle or freeze.

It was the Apollo Lunar Surface Drill, used on Apollo 15, 16, and 17, that eventually extracted core samples from up to three metres below the lunar surface and placed probes for heat-flow experiments. The drill itself was built by Martin Marietta, but its motor - the power head - was built by Black & Decker.

Getting that motor right was the hard part. It had to be small, light, and efficient enough to finish the job on a single battery charge, with no way to plug in for a recharge. To solve it, Black & Decker's engineers developed a computer program that optimised the motor's design for minimum power consumption. That piece of work - not the drill itself - is what mattered for what came next.

Black & Decker had already released the first cordless electric drill in 1961, years before the Apollo partnership. What the NASA work gave them was a deeper understanding of how to build efficient battery-powered motors. In 1979 they introduced the Dustbuster, the small cordless handheld vacuum that has since sold more than 150 million units. Cordless shrub trimmers, grass shears, lightweight cordless medical instruments, and a long line of later consumer tools drew on the same body of motor-efficiency work.

So Apollo didn't invent the cordless drill. What it did do was push Black & Decker to solve a harder problem than they might otherwise have tackled, and the answer they produced is part of why cordless tools today are small, efficient, and common.

Scratch-Resistant Eyewear

Astronaut helmet visors need to be crystal clear and nearly indestructible. That thin barrier between your eyes and the vacuum of space is your lifeline - when your walk outside your spacecraft could mean a simple scratch could impair your vision during critical operations. Out there, where space dust travels faster than a bullet, even a tiny impact could damage your visor.

In the 1970s, researchers at NASA's Ames lab were working on a coating to protect water filtration membranes on spacecraft. The coating had to be durable, optically clear, and ultra-thin.

One of the scientists working on these coatings, Dr. Ted Wydeven, noticed the potential for other transparent surfaces so NASA extended the research to astronaut visors, eventually developing a transparent, abrasion-resistant layer applied using a vacuum-based plasma polymerization technique.

This NASA-derived coating work helped accelerate abrasion-resistant plastic lens technology that was used by the Foster Grant Corporation, the eyewear company, to coat plastic sunglasses and prescription lenses, which had long been prone to scratching.

The scratch-resistant coatings for plastic lenses that we use today were helped along by coating research first pursued for spacecraft and astronaut visors.

Fighting Fire

In space, even fire behaves differently. Without the upward pull of gravity, flames do not rise in the familiar teardrop shape we see on Earth. They hover in spherical clouds and smolder longer at lower temperatures.

This strange behaviour prompted NASA to explore how fires start, spread, and how they could be safely extinguished in microgravity.

It led to a series of experiments conducted aboard the International Space Station, designed to better understand fire dynamics in space.

One of these efforts, known as the Flame Extinguishment Experiment (FLEX), revealed that certain materials can burn more efficiently in microgravity, even at lower temperatures. It also revealed that fires that might seem extinguished can continue to smoulder, reigniting with minimal warning.

These findings have serious implications, not only for astronaut safety, but also for fire management in confined or low-ventilation environments anywhere.

One outcome was the development of fire-resistant textiles capable of withstanding extreme conditions without sacrificing mobility.

These materials included beta cloth, a fire-resistant Teflon-coated glass fiber with blends of aramids and polymers that laid the foundation for what's now common in firefighting suits, race car uniforms, industrial protective gear, and some military uniforms.

It changed how we understand fire suppression in confined, low-oxygen, low-ventilation environments as well as astronaut safety. It helped design new approaches especially where traditional extinguishers might fail including submarines, aircraft cabins, or sealed industrial rooms.

Clean Water from Air

With launch costs running into thousands of dollars per pound, carrying enough water for a long mission is extremely expensive. Instead, the International Space Station relies on the Environmental Control and Life Support System (ECLSS), which includes a water recovery system that recycles wastewater and captures moisture from astronauts' breath and sweat, turning it back into drinkable water. On the ISS, there is very little room for waste, and no natural water supply nearby.

The value of those closed-loop systems does not end in orbit. NASA says space-station water-recycling technology has been licensed and adapted for Earth-based water treatment, including systems used in field and disaster-relief settings. The broader lessons of recycling, filtration, and water recovery have also influenced how engineers think about water efficiency in other difficult environments right here on Earth.

In your own home, newer water-efficient appliances and greywater systems that reuse shower water for garden irrigation use these same closed-loop principles. If you have a high-end water filter, it might even have origins on the ISS.

Staying Warm

Managing temperature is a critical challenge in spacecraft design. In the vacuum of space, the side of a satellite facing the sun can reach scorching temperatures of 250°F (121°C) while the side in shadow plunges to -250°F (-157°C). Without atmosphere to transfer heat through convection, engineers must rely on radiation and conduction to manage these extreme temperature differences.

NASA's work on multi-layer insulation (MLI), those shiny gold or silver blankets you see covering satellites and spacecraft, that we mentioned in Chapter 1, has contributed significantly to advances in household insulation technology.

These space blankets consist of multiple layers of thin, reflective material (often aluminized Mylar) separated by non-conducting spacer material.

The technology found its way into consumer products in several ways. For example, the lightweight emergency thermal blankets in first aid kits are direct descendants of spacecraft insulation. But the influence extends to more common applications in our homes.

Radiant barrier technology, now widely used in attic insulation, draws directly from the principles used in spacecraft thermal control, reducing heat transfer by radiation, making your home more energy-efficient.

Similarly, advances in multi-pane windows with low-emissivity coatings, which reflect infrared radiation while allowing visible light to pass through, benefited from knowledge gained in developing transparent thermal control coatings for spacecraft windows and optical surfaces.

Aerogel, sometimes called "frozen smoke" or "solid air," is unusual crossover from space to home insulation. This material, the lightest solid known, composed of up to 99.8% air, was further developed by NASA for insulating Mars rovers and collecting cosmic dust.

Under NASA contracts, companies like Aspen Aerogels developed flexible, durable aerogel blankets for use in places with wide temperature extremes for spacecraft, Mars rovers, and cryogenic tanks that can cope with deep space cold (near absolute zero) or the heat of atmospheric re-entry (thousands of degrees).

Its insulating properties, up to four times more effective than conventional fiberglass, can now be found in high-performance building insulation and even some clothing items.

Today, aerogel insulation appears in high-performance outdoor gear, ultra-insulated shipping containers, and even building materials and some consumer products. Boots that stay warm in arctic temperatures without bulk and jackets that insulate without weight also owe their effectiveness to material science developed for the unforgiving environment of space.

Camera in Your Pocket

Every time you snap a photo with your smartphone or digital camera, you're benefiting from technology that took a surprising detour through space research. The story of how modern digital imaging developed illustrates how public investment in space exploration can accelerate technological development in unexpected ways.

The Charge-Coupled Device (CCD), the electronic light detector that revolutionized digital imaging, wasn't invented by NASA. It was created in 1969 at Bell Labs by Willard Boyle and George E. Smith, who initially conceived it as a memory device, not an imaging sensor. Its potential for capturing images was quickly recognized, but early CCDs were small, expensive, and suffered from numerous imperfections.

What changed this promising, but limited, technology into the core of modern digital imaging was NASA's insatiable need for better ways to capture images from space.

NASA's Jet Propulsion Laboratory (JPL) and other institutions were seeking better electronic detectors for space telescopes (like the Hubble Space Telescope) and planetary probes. They saw immense potential in CCDs to replace older, less sensitive technologies like photographic plates and vidicon tubes, which had significant limitations for space applications.

Throughout the 1970s and 1980s, NASA invested heavily in research at universities and companies to improve CCD technology. They needed sensors with greater sensitivity (to capture faint, distant objects), lower noise (to distinguish real signals from electronic interference), larger sizes (to capture wider views of the cosmos), and resistance to the radiation environment of space.

Meanwhile, another imaging technology was developing: the complementary metal-oxide semiconductor (CMOS) sensor. Initially, CMOS sensors produced lower-quality images compared to CCDs and were largely dismissed for high-end imaging. However, CMOS had the potential advantage of integrating multiple functions onto a single chip and consuming far less power.

In the early 1990s, Eric Fossum, an engineer at JPL, and his team focused on advancing CMOS image sensor technology. They developed the active pixel sensor (APS), which dramatically improved image quality and performance, creating what Fossum would later call the "camera on a chip." After his work at JPL, Fossum co-founded Photobit Corporation to commercialize the technology, and it quickly began appearing in webcams, industrial inspection systems, and eventually, smartphones.

This is one of those cases where the word "spinoff" understates what actually happened. Today, CMOS sensors are in virtually every camera-equipped device. The low-light photography on your phone, the facial recognition that unlocks it, and even medical imaging systems all rely on this same basic innovation. They are a space technology that became the dominant way the world captures images. If you removed them from modern life, you'd essentially turn the visual internet off.

Simulating Stress: NASTRAN

Back in the mid-1960s, NASA faced the monumental task of designing increasingly complex rockets and spacecraft. They needed a way to predict how structures would respond to the extreme forces of launch, space operations, and reentry without building and testing countless physical prototypes which was a prohibitively expensive and time-consuming process. .

To solve this, NASA commissioned the development of one of the earliest and most sophisticated simulation tools called NASTRAN (NASA Structural Analysis). This software allowed engineers to model complex structures and simulate how they would respond to forces like vibration, pressure, and heat.

It was one of the first large-scale, general-purpose programs for finite element analysis (FEA), a mathematical technique that breaks complex structures into small, manageable pieces to analyze stress, vibration, temperature effects, and more.

After proving invaluable within NASA, the agency released NASTRAN into the public domain in 1971 where it quickly spread through the engineering world. Industries began using it to test structural integrity digitally, before committing to physical builds.

Companies have now developed commercial versions, enhancing and supporting the software for use across industries, and today, derivatives of NASTRAN are industry standards used to design everything from car components and industrial equipment to buildings, bridges, and consumer products.

When you drive across a bridge or board an airplane in the weeks or months ahead, or even undergo joint replacement surgery, you're probably benefiting from stress simulations originally developed for rocket science.

Clean Air from CO_2 Scrubbers

The build-up of carbon dioxide inside a sealed spacecraft can be fatal leading NASA to develop advanced CO_2 scrubbers to capture and neutralize exhaled carbon dioxide from cabin air. These systems had to be compact, lightweight, and reliable over long periods.

Back on Earth, the same filtration principles have influenced thinking about air purification in places where clean air matters most. This includes hospitals, submarines, school classrooms, and even some commercial office buildings.

Although not all a direct line to NASA, the overall work has meant improved air quality in places where breath is shared, and where health depends on it.

Bringing It Home

All of these examples, from the mattress that supports your back, the tool in your garage, the lens on your glasses, the phone in your hand, the air you breathe, reveal a pattern.

Space demanded solutions. And those solutions, once tested in orbit, flowed back into daily life through partnerships, licensing, public policy, and open data.

Each example may seem modest on its own. But taken together, they show how space exploration has functioned as an innovation engine.

Just as space reshaped what we sleep on, breathe, wear, and use, it also reshaped how we care for our bodies.

6

HEALTH AND MEDICINE INSIGHTS FROM ORBIT

Your body is a marvel of adaptation, fine-tuned over thousands of generations to function in Earth's gravity. Take that fundamental force away, and suddenly systems that worked perfectly begin to change.

For space agencies worldwide, keeping astronauts healthy during long missions has become a window into how our bodies work and a catalyst for medical innovations that might be helping you or your loved ones right now.

The Ear Thermometer

Next time someone takes your temperature with one of the small electronic thermometers that you point at a forehead or put in an ear, it's worth knowing where that device came from.

In the 1970s, NASA's Jet Propulsion Laboratory was working on a mission called the Infrared Astronomical Satellite, or IRAS. The problem IRAS was designed to solve is a specific one: you can't measure the temperature of a distant star by touching it with a thermometer. Every object with a temperature above absolute zero emits infrared radiation, and if you can measure the infrared emission precisely, you can calculate the temperature. IRAS's sensors were designed to measure the faint infrared glow of stars, galaxies and nebulae. It launched in 1983 and mapped the entire sky in infrared for the first time.

In the late 1980s, a small San Diego company called Diatek joined JPL's Technology Affiliates Program, looking for ways to adapt space-derived infrared sensor technology to medical use. The principle was the same: an eardrum, like a star, radiates heat, and measuring the radiation tells you the temperature. In 1991, Diatek released the Model 7000 thermometer. It could measure body temperature in less than two seconds, didn't touch the eardrum, and avoided any risk of cross-infection. Compared to the three minutes an electronic oral thermometer took, or the eight minutes for a mercury thermometer, this was transformative for hospitals and paediatric clinics.

The technology has since spread. Infrared forehead thermometers work on the same principle. So do the non-contact thermometers used at airports and workplaces during the Covid-19 pandemic, which let tens of millions of people be screened for fever quickly and without contact.

The same sensor technology sits inside smartphones, in cars for collision avoidance and pedestrian detection, in security cameras, and in firefighters' goggles that let them see through smoke.

In each case, the underlying ability to detect and interpret infrared radiation at useful precision and acceptable cost exists partly because space astronomy needed it first.

Bones, Muscles and Gravity

In space, astronauts can lose bone density faster than a person with advanced osteoporosis. They can lose 1-2% of their bone density each month in space, particularly in weight-bearing bones like the spine, hips, and legs. That's roughly 10 times quicker than what an elderly person with osteoporosis might experience.

And yet, they're often in peak physical health before launch. That contrast of healthy bodies weakening under the wrong conditions has been a revelation. It's helped scientists understand not just how bone degrades, but why.

The culprit is weightlessness itself. On Earth, our bones constantly rebuild themselves in response to the stress of gravity and movement. Every step you take sends signals to your bones that they need to maintain their strength. In microgravity, those signals disappear, and bones begin releasing calcium and other minerals at an accelerated rate.

In other words, our bones depend on loading to stay strong, and space gives researchers a uniquely clear way to study what happens when that loading disappears.

To slow this loss, NASA developed the Advanced Resistive Exercise Device (ARED), a sophisticated machine that mimics weightlifting in a weightless environment using vacuum cylinders. It allows astronauts to load their muscles and bones with carefully measured resistance, maintaining strength and density through targeted exercise.

As well as these benefits, the medical imaging techniques used to monitor bone changes in orbit, like quantitative computed tomography (qCT), have contributed to wider thinking about how clinicians assess skeletal weakening, helping to spot early signs of weakening before a fracture ever occurs.

Additionally, the equipment, including the resistance training systems, fine-tuned muscle sensors, and real-time strength tracking built for astronauts, has influenced rehabilitation thinking and exercise design around the world.

The same data that supports astronauts has informed clinical guidance: which movements are most effective, what level of resistance is needed, and how often it must be applied.

What began as cutting-edge hardware on the International Space Station has helped shape rehabilitation thinking and exercise design for people recovering from injury, surgery, or struggling with brittle bones. Such specialized protocols have proven particularly valuable for patients with severe osteoporosis, where traditional exercise might actually increase fracture risk.

The precise, controlled resistance techniques developed for astronauts offer a model for safer, more controlled resistance work. And the accelerated aging process we have talked about in this section may be where it has had the biggest impact.

By compressing years of bone loss into months, space research provided a unique laboratory for studying treatments.

When we see bone density decreasing 10 times faster than normal, researchers can evaluate potential treatments in a fraction of the time it would take on Earth. This has given researchers a model for studying bone loss and possible countermeasures for hundreds of millions of people worldwide suffering from osteoporosis, particularly benefiting women after menopause who face similar challenges for different biological reasons.

Muscle deterioration follows a similar pattern. On Earth, loss of muscle mass, especially postural muscles, is a slow decline that can leave older adults prone to falls and long recoveries. In the microgravity environment of space, that same process is visible within days.

Europe's contribution, the Muscle Atrophy Research and Exercise System (MARES), allowed scientists to isolate individual muscle groups, tracking strength, speed, and fatigue under highly controlled conditions. This has helped researchers study muscle decline with a precision that can inform rehabilitation and sarcopenia research.

Through the pathways we'll explore later, some of these findings and technologies have moved from specialized space applications into wider healthcare research and practice.

The knowledge gathered from our astronauts is now helping millions of people stay stronger, more mobile, and more independent for longer. Yet the benefits extend far beyond our bone and muscle health.

Robotic Arms

On the Space Shuttle, a robotic arm was needed to capture satellites, position cargo, and carry out delicate operations in orbit.

The best-known result was Canadarm, followed later by Canadarm2 and Dextre. The engineering lessons from those systems did not stay in space.

In Canada, the same robotics expertise helped shape neuroArm, a surgical robot developed by MDA and a team led by Dr. Garnette Sutherland at the University of Calgary. Designed to operate inside an MRI scanner, neuroArm combined precise remote control with haptic feedback, allowing surgeons to feel tissue through the machine.

In this case, the connection between space robotics and medical robotics is direct rather than metaphorical and although neuroArm did not create the whole field of surgical robotics, it is a clear example of how robotics developed for space operations helped inform high-precision medical systems on Earth.

Your Heart in Space

We have all done it at some point - stood up too quickly and felt dizzy. That momentary lightheadedness happens because your cardiovascular system needs a moment to adjust to gravity pulling blood away from your head and toward your feet.

Now imagine the opposite effect happening for months at a time. In the absence of gravity, where there is no force to pull fluids downward, astronauts experience what's called a "headward fluid shift."

Blood and other fluids that would normally pool slightly in the lower body move upward instead, giving astronauts a characteristic "puffy face" and "bird legs" appearance.

The heart works harder to handle the increased fluid volume in the upper body and, over time, it might even shrink slightly because it doesn't have to pump as forcefully - it is not fighting against gravity.

It means that when astronauts return to Earth, they often experience "orthostatic intolerance" which results in dizziness or even fainting when standing up, as their cardiovascular system struggles to readjust to gravity's pull.

This cardiovascular gravity-related problem provides a great example of how space can unwittingly create an accelerated model of conditions that we can see on Earth.

This is because fluid shifts and heart adaptations that happen over days or weeks in orbit might take months or years to develop in patients with certain heart conditions on Earth where gravity slows down the symptoms. Just as with the bone density research we discussed earlier, this compression of time allows researchers to study interventions more efficiently.

Advanced ultrasound techniques and wearable sensors developed for space missions have added to our understanding of how the heart and circulation respond to changing conditions.

The Japan Aerospace Exploration Agency (JAXA) has conducted particularly detailed studies on cardiovascular changes in microgravity, complementing NASA's research through the kind of international collaboration we touch on again later.

The miniaturized ultrasound systems designed to work in the confined space of a spacecraft have helped accelerate the use of portable ultrasound and remote-guided scanning in emergency rooms and rural clinics, and even disaster zones.

And the wearable sensors that track astronauts' heart function continuously during space missions have influenced the development of cardiac monitoring systems that patients can wear at home, allowing doctors to detect irregularities earlier and more accurately than ever before.

POTS

Postural Orthostatic Tachycardia Syndrome (POTS), in which standing can trigger an abnormal rise in heart rate along with dizziness and fatigue, has been better understood through this research.

It affects an estimated 1-3 million Americans, predominantly women, and the strategies explored for astronauts, including exercise protocols and compression garments, have also helped inform how it is managed.

Related approaches are also used in managing other conditions affecting blood pressure regulation, from chronic orthostatic hypotension to recovery after prolonged bed rest.

Heart Failure

The miniaturized wireless sensing work developed for astronaut vital-sign monitoring has also found its way into heart care.

Engineers at NASA's Glenn Research Center created a tiny radio-frequency telemetry system to reduce the need for cumbersome wired sensors during astronaut exercise. Licensed for cardiovascular use, that technology later helped lead to the Cordella Pulmonary Artery Sensor and Heart Failure System, an implantable monitor that can detect rising pulmonary artery pressure before a heart-failure crisis becomes severe.

Used in this way, it can support earlier medication adjustments and has been associated with fewer hospitalizations.

Microgravity research has also contributed in a different way. By studying how fluids shift through the body in space, researchers have gained insights into circulation and pressure regulation that are relevant to conditions such as heart failure, where fluid balance is often critical.

The Heart Pump

Perhaps the most surprising cardiovascular innovation we've seen from space research is the DeBakey VAD (Ventricular Assist Device).

This device, which helps failing hearts pump blood effectively, emerged from a collaboration between NASA engineers and renowned heart surgeon Dr. Michael DeBakey.

NASA's expertise in creating durable, efficient pumps for spacecraft systems was applied to develop a miniaturized heart pump based on space shuttle fuel pump concepts.

The result is a device that is smaller, more efficient, and more durable than previous designs, allowing patients with severe heart failure to live longer, and have more active lives while waiting for transplants or even as a permanent solution.

Since receiving FDA approval, this technology has saved many lives and shows how the precision engineering required for space missions can translate directly to life-saving medical devices.

These are only some of the ways space research has contributed to heart care.

Telemedicine's Space Origins

Long before the COVID-19 pandemic made telehealth mainstream, astronauts were being monitored and treated from hundreds of miles away. NASA and other space agencies had to pioneer ways to assess health, diagnose problems, and even guide treatments without a doctor physically present.

Those early biomedical monitoring systems helped establish practical models for remote monitoring over distance.

Those early biomedical monitoring systems used during the Mercury and Apollo missions - compact, reliable devices that could transmit vital signs across vast distances - helped establish practical models for remote monitoring over distance.

Combined with data compression algorithms and transmission protocols it has all helped to make modern telehealth possible.

It again illustrates the "necessity drives innovation" principle we have explored, where the extreme constraints of space force solutions which can eventually benefit everyone.

Remote Care

Perhaps even more valuable were the operational protocols developed for remote diagnosis and consultation between astronauts and flight surgeons. NASA's experience with the Integrated Medical Group Operations (IMedO), which coordinates all health services for astronauts on the International Space Station, created a blueprint for effective remote healthcare delivery.

An incredibly important development came through NASA's work with ultrasound technology. On Earth, ultrasound typically requires a trained sonographer and a physician to interpret the results. But in space, astronauts, who aren't medical specialists, needed to conduct these examinations themselves with guidance from doctors on the ground.

NASA developed specialized protocols and training systems that allowed non-medical personnel to perform complex ultrasound examinations with remote expert guidance.

Subsequently, healthcare workers without specialized training can conduct ultrasound exams while experts provide real-time guidance from hundreds of miles away in rural and underserved areas.

Remote Medicine and Cancer Detection

Sometimes the largest impacts of space technology on our lives come from unexpected transfers of knowledge between seemingly unrelated fields. Medical imaging provides a good example of this cross-pollination.

As we know, in the 1960s, engineers at NASA's Jet Propulsion Laboratory developed digital image-processing techniques to sharpen and correct spacecraft pictures returned from missions to the Moon and planets. This helped Robert Nathan and colleagues adapt new methods for medical imaging, including x-ray photographs.

Over time, advances in digital enhancement, contrast handling, and noise reduction became part of the wider evolution of medical imaging, including cancer detection.

This is a great example of how NASA didn't invent medical imaging itself, but its contributions - especially in image processing - improved and expanded its capabilities.

Modern systems now routinely use advanced image processing techniques, some drawing on the methods advanced through space imaging, helping radiologists to identify potential tumors at earlier, more treatable stages.

Similarly, image analysis techniques first developed to interpret Landsat satellite data, such as multispectral imaging and digital pattern recognition, have been adapted for use in dermatology. These methods help doctors analyze photographs of skin lesions, distinguishing between benign moles and potentially dangerous melanomas by detecting subtle color and pattern differences that might otherwise go unnoticed.

These image processing techniques, often shared through NASA's Technology Transfer Program, have been incorporated into much of medical imaging equipment used around the world. An MRI of a knee injury, a screening for breast cancer, or a CT scan checking for internal injuries might all have benefited from this work.

Growing Space Plants to Treating Wounds

Red light therapy, now widely used for pain relief and skin treatments, has an unexpected origin story that begins not in a medical lab, but in NASA's quest to grow food in space. We cover food itself in more detail in Chapter 8.

In the late 1990s, NASA scientists at the Marshall Space Flight Center were experimenting with light-emitting diodes (LEDs) to grow plants aboard the Space Shuttle. Plants need light for photosynthesis, but traditional growing lights generated too much heat and consumed excessive power for spacecraft use. LEDs offered a more efficient alternative that could be precisely tuned to the specific wavelengths plants need most.

During this research, scientists made a surprising discovery: certain wavelengths of light, particularly red (660nm) and near-infrared (850nm) light emitted by LEDs, could penetrate human tissue up to several centimeters deep and stimulate cellular activity.

This light energy seemed to quicken wound healing and reduce pain and inflammation by boosting mitochondrial function, the cellular power plants that drive our body's energy production.

The implications were significant for both astronauts and people on Earth. In space, where minor injuries heal more slowly due to microgravity effects, such technology could be crucial. On Earth, it offered new hope for conditions that resist conventional treatment.

NASA recognized the potential and partnered with medical

researchers to develop this technology further, eventually patenting several applications.

The Quantum Devices company, working with NASA through the Technology Transfer Program, developed the WARP 10 (Warfighter Accelerated Recovery by Photobiomodulation), a hand-held LED unit first used by Navy SEALs for combat wound treatment.

These devices have been used in treating slow-healing diabetic ulcers that affect millions of people worldwide, manage pain from arthritis or muscle injuries, reduce inflammation after surgery, and even address certain applications in skin rejuvenation and hair restoration.

Clinical studies have shown that this light therapy can accelerate wound healing by up to 40% in some cases and significantly reduce the need for pain medication in others. For diabetic patients at risk of amputation due to non-healing ulcers, this space-derived technology has been used as part of wound-care strategies aimed at reducing severe complications.

As strange as it sounds, if you notice someone using a red light therapy panel for pain relief or a medical professional treating a wound with a specialized light device, you're seeing one of the clearest medical applications to emerge from technology first explored during NASA's work on growing food in space.

Space Technology in Neonatal Care

One later application of related monitoring and nutrition research can be seen in neonatal care. Like astronauts, premature babies must adapt to environments very different from those their bodies were prepared for. Both require constant monitoring of vital signs. And for both, even small fluctuations in temperature, oxygen levels, or fluid balance can have serious consequences.

Traditional monitoring required attaching multiple wires to an infant's delicate skin - limiting movement, complicating care, and increasing infection risk.

Wireless, non-invasive monitoring systems now allow comprehensive tracking of heart rate, respiration, oxygen saturation, and temperature with minimal contact, reducing skin damage and allowing for more natural positioning and movement - factors known to be crucial for neurological development.

Even nutrition has seen space-to-NICU technology transfer.

Research into algae-based life support systems for long-duration space missions led to the discovery of methods to produce docosahexaenoic acid (DHA), an omega-3 fatty acid critical for brain and eye development.

NASA-backed research helped advance algae-derived DHA now used in infant formula, and we talk about this in the chapter From Space to Plate.

The telemedicine systems already discussed have particular relevance in neonatal care, where specialists are often concentrated in large medical centers. Remote monitoring and consultation systems allow neonatologists to provide expert guidance to community hospitals, ensuring premature infants receive specialized care even when born far from major medical centers.

During the COVID-19 pandemic, these systems became even more crucial, allowing parents to see their babies when visitation was restricted.

A Universe of Health Innovations

From understanding how bones maintain their strength to developing telemedicine systems that bring specialized care to remote areas, from enhancing medical imaging to discovering new ways to promote healing, space exploration continues to provide surprising benefits for healthcare on Earth.

These innovations reach us through various pathways, some through scientific publications that advance our fundamental understanding of human physiology, others through patents licensed to medical device companies, and still others through the transfer of expertise as engineers and scientists move between the aerospace and healthcare sectors.

The DeBakey VAD heart pump and the WARP 10 light therapy device are examples that show how solving problems for space travel can create solutions for human health.

The extreme constraints of space, with limited resources, isolation from help, and harsh conditions, force inventions that must work reliably the first time. This leads to solutions that are often more elegant, efficient, and effective than those developed under less demanding circumstances.

It means that if you have a bone density scan, use a virtual doctor's appointment, undergo minimally invasive surgery, or receive treatment with advanced imaging guidance, you are witnessing how our desire to explore space can directly help us individually in unexpected ways.

7

SLEEP, STRESS, AND THE SPACE MIND

In space, clocks lose their natural anchor. Outside, the Sun rises and sets every 90 minutes, and the body, evolved for a steady 24-hour rhythm, begins to lose its bearings.

Add isolation, confinement, and the subtle strain of living in a place where a loose bolt can mean catastrophe, and you get a recipe for mental and physiological stress unlike any on Earth.

This is the setting that has driven a revolution in our understanding of sleep and mental health.

It has meant that NASA and other space agencies have spent decades trying to answer the question of how to help humans maintain psychological balance and healthy sleep when the environment undermines every cue the body normally relies on.

In pursuing that question, space scientists have helped advance research into sleep, circadian regulation, and behavioural support tools, some of which may also have uses here on Earth.

The Body Clock

One of the hardest parts of space travel isn't mechanical but biological. Humans are deeply circadian creatures and our core body temperature, hormone levels, attention span, mood, and sleep cycles all follow a roughly 24-hour rhythm.

This internal clock, set by exposure to natural light, becomes scrambled in orbit. Astronauts aboard the International Space Station experience 16 sunrises a day. Without Earth's consistent day-night cycle, the circadian system becomes confused, drifting out of sync with external time cues and leading to disrupted sleep, reduced cognitive performance, and mood instability.

Researchers observed shifts in astronauts' core body temperature rhythms during missions, indicating circadian imbalance. But understanding the mechanism wasn't enough, they needed a way to fix it.

The solution came in the form of dynamic lighting systems that mimic the progression of natural light on Earth. Related dynamic-lighting approaches are now being used or tested in settings such as hospitals and care environments including dementia wards, helping patients maintain a more stable sleep-wake cycle in environments where windows are absent or meaningless.

But understanding when we sleep is only half the puzzle. The next challenge was figuring out how to measure that sleep without interfering with it.

Tracking Sleep

In space, fatigue is not just a comfort issue. It can affect reaction time, judgment, memory, and attention at exactly the moments when crews may be handling complex equipment or responding to unexpected problems. That is one reason sleep research became such a serious part of mission planning. In an environment where the margin for error is small, rest is not a luxury. It is part of safety.

Early sleep studies in space were clunky affairs, involving cumbersome EEG setups that interfered with natural rest. But the need to track sleep efficiently in weightless, low-resource environments led to innovation.

Mission planners use lighting schedules, workload timing, and carefully structured daily routines to help crews keep a stable rhythm. In some cases, astronauts have also used sleep medications, though this raises its own concerns about dependence, alertness on waking, and the limits of trying to medicate an environmental problem.

Dry-EEG headbands and, more recently, ear-EEG sensors are now being developed to monitor sleep architecture non-invasively. These devices should be able to capture brainwave activity without the need for conductive gels or full lab setups.

And related portable sleep-monitoring tools are now part of a broader move toward less intrusive sleep assessment outside traditional lab settings.

Some of these solutions are also being explored in clinical sleep assessment and circadian research to help guide treatment. The original impetus may have been to understand sleep quality and disruption in astronauts, but in the end it meant a broader understanding of sleep health, especially valuable in communities without easy access to specialist care.

Even with better tools to monitor sleep, rest remains elusive when the mind is under pressure. It's not just physiology that strains under stress but the whole psychology of being alone together.

Psychology of Isolation

The psychological challenges of space are not dissimilar to those experienced in remote research stations, submarines, or during global lockdowns. But in space, they're amplified. Astronauts endure long periods away from loved ones, tight quarters with limited social contact, and a near-total absence of unstructured solitude. The stakes for psychological well-being are existential, not just personal.

Spaceflight research has also encouraged an interest in automated psychological support tools, especially for future missions where communication delays could make real-time contact with clinicians impossible. These ideas might overlap with wider developments in digital mental health, including guided self-help and CBT-based apps.

They are not replacements for human clinicians, but they reflect a wider move toward guided self-help and structured digital support.

Emotional support systems helped, but the need for mental escape remained. Without forests, beaches, or skies, astronauts turned to technology to reimagine the idea of space altogether.

The VR Forest

When your physical world is confined to a few pressurized compartments, the mind needs somewhere else to go. This is the idea behind virtual reality therapy for spaceflight, which can create immersive environments that simulate peaceful, open, nature-filled spaces such as a stroll through a forest.

These simulations are designed to calm the nervous system and, in some versions, link visual cues to breathing or heart rate. Back on Earth, VR-based relaxation and stress-management tools are also being explored in a range of therapeutic and wellbeing settings.

Sleep and Sanity

A valuable insight from this field is the recognition that sleep and mental health are not separate domains.

In space, disturbed sleep correlates with increased emotional reactivity and impaired decision-making. On Earth, the link is just as strong. Poor sleep quality contributes to depression, anxiety, and impaired immune function. Space research has reinforced the idea that sleep, stress, and emotional regulation are tightly connected, a view that also informs wider health research.

Together, these lines of research have helped improve our understanding, showing that resilience is a system rather than a single trait. Rest is not a luxury, but a baseline for functioning. They also suggest that even in highly artificial environments, cues that echo nature, whether through VR or carefully timed light, may help support mental balance.

Reentry

When astronauts return from their mission, they face a different kind of disorientation. The body has to re-adapt to gravity, the mind to social complexity, and the sleep-wake cycle to natural light.

This process, known as re-entrainment - the body's way of resyncing its internal clock to external cues - is closely monitored by space agencies. But it also mirrors the challenges that people face when changing time zones, leaving hospital care, or returning to civilian life after military deployment.

Re-entrainment in astronauts also mirrors the slower recovery people face after shift changes, long-distance travel, or prolonged isolation and studying this re-entrainment in astronauts can help our understanding of similar recovery problems that some of us live with daily.

But survival also depends on what feeds the body. And in space, even food becomes a question of science, innovation, and adaptation.

8

FROM SPACE TO PLATE

In previous chapters we looked at what space gave us in medicine, in sleep research, in the products and technologies most of us use without thinking twice. Food is a less obvious entry in that list, but the problems of feeding people in space turned out to be so extreme, and so specific, that solving them pushed food science in directions that eventually reached your kitchen.

The Early Years: Mercury and Gemini (1961-1966)

When Yuri Gagarin became the first human to eat in space in 1961, his meal, pureed meat from a toothpaste-like tube, was hardly appetizing by today's standards. If you've ever sipped applesauce through a pouch on a hike or handed one to a toddler, you've got a taste of what John Glenn ate a few months later aboard Mercury.

The first American space meals were designed with the primary goal of basic survival. Engineers weren't concerned with taste, texture, or overall enjoyment; they simply needed to ensure astronauts received adequate nutrition during their brief missions.

Mercury astronauts subsisted primarily on tube foods, freeze-dried powders, and compressed food cubes coated with gelatin to prevent crumbling. Water was essential for rehydrating the dried foods and also for drinking, adding complexity to the mission's logistics including its impact on the weight that needed to be carried.

By the time the Gemini program launched in 1965, food options had improved somewhat. The menu expanded to include freeze-dried foods that could be rehydrated with water, providing astronauts with shrimp cocktail, chicken and vegetables, butterscotch pudding, and applesauce.

While still limited, these meals represented the first step toward making space food more palatable and the Gemini astronauts like John Young and Gus Grissom also experimented with new packaging and preparation methods to make eating in zero-gravity more manageable.

Gemini also brought a small but meaningful change: spoon-bowl packages that let astronauts eat with something closer to a normal utensil. That may not sound revolutionary, but imagine the comfort of scooping a warm meal compared to squeezing it from a tube.

Apollo Era: The Birth of Modern Food Technology (1967-1975)

With missions lasting up to two weeks and venturing further from Earth than ever before, the Apollo program brought new challenges and the food needed to be more varied and satisfying while still meeting strict safety requirements.

The Apollo era also saw the development of other technologies that would later change food preservation. The need for lightweight, shelf-stable, and safe food drove innovations in freeze-drying, thermostabilization, and packaging that would eventually find their way into commercial products. The Apollo missions required extensive collaboration between food scientists, engineers, and astronauts, leading to a much more comprehensive approach to food development.

This era saw the introduction of the "wetpack," thermostabilized, ready-to-eat foods in aluminum cans that didn't require rehydration, and in 1968, Apollo 8 featured the first Christmas dinner in space that included thermostabilized turkey and gravy, marking a festive milestone in space dining.

But the biggest food safety advance from this era was a food safety process called Hazard Analysis and Critical Control Points (HACCP) which we cover in detail in the following pages.

Skylab: The First Space Kitchen (1973-1979)

Skylab, America's first space station, represented a huge leap in space food systems. For the first time, astronauts had access to a refrigerator, a freezer, and a food-warming tray. The 8-foot-diameter dining room table even had foot restraints, allowing crews to sit down together for meals - a social experience previously impossible in space. This improved dining environment played a significant role in boosting the astronauts' morale and overall psychological well-being.

The Skylab menu featured over 70 different food items, including ice cream. Astronauts could choose their meals before the mission, adding a welcome element of personal preference. The improved food system contributed to the success of longer-duration missions, with crews staying in space for up to 84 days. Skylab's food system even allowed for individual dietary needs and preferences.

Skylab also saw the first significant research into the nutritional needs of humans living in space for extended periods. Scientists discovered that astronauts' bodies processed nutrients differently in microgravity, leading to new insights about human metabolism that would later inform nutritional science. Meanwhile, the studies conducted on Skylab revealed the importance of vitamins, minerals, and trace elements in maintaining bone density and muscle mass in a weightless environment.

The Shuttle Era: Normalizing Space Dining (1981-2011)

The Space Shuttle program brought even more refinements, with improved packaging, greater variety, and the introduction of the now-famous "space tortilla" - a flatbread that solved the problem of bread crumbs floating in microgravity.

Shuttle astronauts enjoyed a rotating menu of over 200 items, including many recognizable brands that had been specially packaged for space. It even included food for special diets, international cuisines, as well as the occasional treat like M&Ms or cookies.

The Shuttle era also saw the development of the Extended Duration Orbiter (EDO) food system for longer missions. These innovations would later influence commercial ready-to-eat meals and camping foods, as well as changes to food packaging that reduced waste and improved convenience.

International Space Station: Global Cuisine in Orbit (2000-Present)

Today's International Space Station (ISS) food system is the culmination of decades of research and development. Astronauts from different countries bring their national cuisines to orbit, creating a truly international dining experience.

It has developed to such an extent that it combines nutrition along with enjoyment and psychological well-being. Astronauts can request favorite items, and the ISS crew members can also take part in meal planning and preparation including the food needed for regular celebrations across different cultures - all of which helps to keep a positive and supportive atmosphere on board.

The ISS has also become a laboratory for food technology research, with experiments on growing fresh produce in microgravity.

The Vegetable Production System, nicknamed "Veggie," has successfully grown lettuce, cabbage, kale, and zinnia flowers in orbit. In 2015, astronauts harvested and ate space-grown lettuce for the first time, a small but significant step toward sustainable food production for future long-duration missions.

The Veggie system has also been used to study the effects of microgravity on plant growth and development, leading to new insights into plant biology.

Mars and Beyond

As space agencies prepare for potential missions to Mars, food systems face bigger challenges. A Mars mission could last three years or more, requiring food that remains safe, nutritious, and appealing over a very long period of time. The food systems will need to be highly efficient and sustainable to minimize the need for resupply and it is this that has spurred research into closed-loop systems that recycle water and nutrients.

NASA's Advanced Food Technology Project is now developing solutions for these future missions, including shelf-stable foods with a 5-year shelf life, bioregenerative food systems that can produce fresh food during the mission, automated food production systems that require minimal crew time, and even research into 3D food printing for customized meals.

As we'll see throughout the rest of this chapter, the solutions already developed for feeding astronauts continue to influence how we produce, preserve, and consume food here, on our home planet.

Keeping Food Safe and Nutritious

When you tear open a packet of freeze-dried strawberries during a hike or heat up a ready-to-eat meal from a flexible pouch, you're benefiting from preservation technologies that were perfected for space travel.

The challenge of feeding astronauts safely in space did more than change menus. NASA's unusually strict food requirements helped drive improvements in food processing, preservation, sterilization, packaging, and nutrition. Space food had to remain safe over long periods, resist contamination, retain nutritional value, and work in an environment where refrigeration, resupply, and medical backup were limited. Those demands pushed food scientists and engineers toward methods that later proved useful far beyond the space program, in industry, emergency food systems, and consumer products.

Freeze-Drying

Freeze-drying existed before the space-age, but NASA's need for lightweight, shelf-stable foods helped push it into wider practical and commercial use.

For space missions, reducing weight while preserving nutrition and usability mattered enormously, and freeze-drying offered a practical solution.

The process involves freezing food and then removing the ice through sublimation (converting it directly from solid to vapor without passing through a liquid state). This process preserves the food while maintaining its structural integrity.

What makes freeze-dried foods particularly valuable is their combination of light weight (it reduces the weight of food by up to 80%), long shelf life (up to 25 years for some products), and nutritional retention.

Unlike traditional dehydration, freeze-drying preserves much of the food's original flavor, texture, and nutrients because it doesn't expose food to high temperatures. This also helps maintain the color and aroma of the food, enhancing its palatability.

Using its experience in spacecraft food systems, NASA helped develop shelf-stable foods for homebound elderly people, and that work in turn prompted companies to manufacture compressed and freeze-dried foods for campers and compact emergency food rations.

Flexible Packaging Revolution

If you've ever heated up ready-to-eat rice in a microwave or enjoyed a shelf-stable curry from a flexible pouch, you've benefited from another food-packaging format that was adopted early and refined through space-food systems.

NASA's need for lightweight, compact, and shelf-stable food packaging for space missions helped accelerate the development and wider use of flexible, shelf-stable pouch packaging for food.

Commercial firms began producing astronaut-type meals for public distribution in retort pouches, an innovative flexible package that combined some of the advantages of metal cans and boil-in-bag foods while helping block oxygen and moisture.

The technology was first used extensively during the Apollo program, where retort pouches allowed astronauts to enjoy a greater variety and quality of meals in space.

Retort pouches are now widely used in ready-to-eat food packaging, including meals, baby foods, and military rations.

Food Safety Without Chemicals

Food irradiation is another preservation method that gained importance in the space-food context. It was not invented for NASA, but the need to provide astronauts with food that was exceptionally safe and stable over long missions made irradiation an attractive option.

By using ionizing radiation to reduce bacteria, parasites, and other contaminants, food irradiation offered a way to improve safety without relying on refrigeration.

NASA studied and used irradiated foods as part of its wider effort to solve the twin problems of contamination control and shelf stability in space.

Irradiation has since become one of several approved methods used to improve food safety and extend shelf life in selected products.

While consumer acceptance has been slower than for some other preservation methods, it remains valuable in situations where safety and shelf life are especially important.

More broadly, space-food research ran beside wider advances in sterilization, contamination control, and shelf-stable packaging, all of which became increasingly important as our food systems became more complex.

HACCP: The Space-Age Food Safety Revolution

In the early 1960s, as NASA prepared for longer space missions, they faced a critical challenge: how to ensure absolute food safety for astronauts. A foodborne illness in space could be catastrophic, potentially compromising the entire mission and endangering lives with no access to medical facilities.

To address this challenge, as mentioned earlier, NASA partnered with the Pillsbury Company to develop a revolutionary approach to food safety. The work created the Hazard Analysis and Critical Control Points (HACCP) system, a preventive approach that would transform food safety worldwide.

This systematic, preventive approach to food safety was designed to prevent contamination rather than simply testing the final product after it had been manufactured.

Dr. Howard Bauman, Pillsbury's lead scientist on the project explained that they couldn't just test for problems after the food was produced, they had to build safety into the entire process from the beginning, identifying potential hazards and controlling them at every step.

The HACCP approach represented a fundamental change in thinking about food safety. Rather than focusing on inspecting the final product, it identified potential hazards at every stage of food production, establishing critical control points where those hazards could be prevented, eliminated, or reduced to safe levels.

For the Apollo missions, this meant analyzing every step in the food production process, from raw ingredients to processing, packaging, storage, and preparation, to identify where contamination might occur and implementing controls to prevent it.

The system proved so effective that Pillsbury soon implemented it for their commercial food production. In 1971, after a national crisis involving glass contamination in baby food, the FDA began incorporating HACCP principles into its food safety regulations.

Today, HACCP is the global standard for food safety, required by regulatory agencies worldwide for everything from meat and seafood processing to juice production and restaurant operations. The system NASA developed to protect a handful of astronauts now safeguards billions of people every year.

Frank Yiannas, former FDA Deputy Commissioner for Food Policy and Response, has described HACCP as the foundation of modern food safety management and often highlights its origins in the space program. (*Yiannas, F. (2015). Food Safety = Behavior.*)

When you enjoy a meal without getting sick, whether at a fast-food restaurant, a fine dining establishment, or your own dinner table, you can thank HACCP.

Nutrition Under Pressure

In space, food has to do more than stop hunger. It has to help protect bone, muscle, immunity, and performance over time. That has made nutrition an important part of mission planning. NASA's food and nutrition teams study how nutrients behave during long storage, how diets need to change in microgravity, and how food systems can support health on missions where resupply is limited and the body is under unusual strain.

That challenge has pushed food science beyond calories and convenience. Space crews need meals that remain safe, palatable, and nutritionally reliable for months or even years. They also need those foods to help counter problems that show up in orbit, including bone loss, muscle wasting, altered appetite, and stress on the immune system. In that sense, food becomes part of the medical and operational toolkit, not just a source of energy.

This does not mean space agencies invented modern nutrition science. It means long missions forced researchers to ask unusually practical questions, such as which nutrients remain stable over time, which are lost during storage, and which matter most when the body is adapting to microgravity and confinement. Those questions are important more generally too, especially in situations involving aging, illness, recovery, isolation, or limited access to fresh food.

Seen this way, space food is not only about engineering meals that survive launch and storage. It is also about understanding how food can help the body stay resilient when conditions are working against it. That way of thinking has influenced much of NASA's food and nutrition research and helps explain why space science has had such a lasting influence on how we think about nourishment under pressure.

Calcium and Vitamin D

Studies on the ISS have shown that vitamin D supplementation, combined with calcium intake and the resistance exercises we mentioned in Chapter 6, can significantly reduce bone loss.

This research adds to wider understanding of how calcium, vitamin D, and exercise help protect bone health during long periods of reduced loading.

It has informed recommendations for older adults, particularly post-menopausal women at risk for osteoporosis and the specific combinations and dosages developed for astronauts have influenced clinical guidelines for preventing age-related bone loss.

The space program's emphasis on bioavailable forms of these nutrients, those most easily absorbed and used by the body, has also shaped supplement make-up.

Protein and Amino Acids

Muscle atrophy is the other side of the same problem. Here NASA's research has focused on protein quality, quantity, and timing to preserve lean muscle mass during long-duration trips.

Their studies identified specific amino acid profiles, particularly branched-chain amino acids (BCAAs), that help maintain muscle even during periods of reduced physical activity.

This research adds to wider understanding of muscle maintenance during inactivity and aging including athletic nutrition, rehabilitation medicine, and geriatric care. Sports nutrition products featuring optimal protein timing and specific amino acid ratios are often linked to this research and many medical nutrition products for bedridden patients incorporate these findings to help prevent muscle wasting during periods of inactivity.

Omega-3 Fatty Acids

A direct example of how space nutrition research can benefit consumers comes from studies on omega-3 fatty acids, particularly docosahexaenoic acid (DHA).

In the 1980s NASA scientists investigated algae-based sources of DHA to support astronauts' brain and eye health during long missions, as these tissues are particularly vulnerable to oxidative stress in the space environment.

This innovation provided a vegetarian, sustainable alternative to fish-based DHA, and the ingredient is now used in human nutrition products, including infant formula, prenatal vitamins, and dietary supplements.

The inclusion of DHA and ARA (another fatty acid) in infant formulas was helped along by this line of research. Today, over 90% of infant formulas sold in the US contain algae-derived DHA, included for its potential role in brain and visual development.

Antioxidants

Space is an environment of heightened oxidative stress, so NASA has also studied nutrients that may help protect cells under pressure.

NASA's research into protective nutrients, including antioxidants such as vitamin E, vitamin C, and selenium, sits alongside wider research into how diet supports resilience, recovery, and long-term health

Prebiotics and Probiotics

Recent research on the ISS has focused on the gut microbiome, the community of beneficial bacteria that support digestion, immunity, and even mental health. Studies have shown that the space environment can disrupt this delicate ecosystem, potentially compromising astronaut health.

NASA's investigations into prebiotic fibers and probiotic supplements to support microbiome health have paralleled growing commercial interest in these products.

The research has helped identify specific bacterial strains and fiber types that remain stable during long-term storage and effectively colonize the gut even during periods of physical and psychological stress.

The ISS work adds to broader interest in how isolation, stress, and diet affect the gut microbiome.

The Future of Food

A forward-looking aspect of space nutrition research is the move toward personalized nutrition.

NASA scientists recognize that individual astronauts respond differently to the space environment based on their genetic makeup, preflight condition, and other factors. This has led to research on tailoring nutritional countermeasures to individual needs, work that parallels the emerging field of nutrigenomics.

As we prepare for missions to Mars and even further afield, the nutritional challenges will become even more complex, and the potential benefits will become even greater.

As we prepare for missions to Mars and as missions grow longer and more ambitious, food will remain one of the most practical and revealing parts of space research. The same questions raised in orbit, about storage, resilience, nutrition, and waste, continue to matter on Earth as well. The food science that began with tubes of pureed meat will face its greatest test yet.

9

SPACE PARTNERSHIPS

For most of the 20th century, space exploration was the exclusive domain of government agencies and they gave us those iconic images burned into our collective memory from Neil Armstrong's first steps on the Moon, the Voyager probes sailing past giant planets, to the Hubble Space Telescope peering into the depths of the cosmos.

But recently, different names have begun dominating the headlines such as SpaceX, Blue Origin, Rocket Lab, Planet, Maxar, Virgin Galactic, Astrobotic. A new cast of characters has joined humanity's space adventure - private companies, often led by visionary entrepreneurs with deep pockets and even deeper ambitions. They're building rockets, launching satellites, planning space stations, and even setting their sights on the Moon and Mars.

And this is one of those occasions where two things are true at once. Private space companies are doing things public agencies couldn't or didn't. They are lowering launch costs dramatically, bringing internet access to places government programs never reached. And public agencies continue to do things private companies don't. Things like long-term fundamental research, global data gathered and released as a public good, technology deliberately designed to spread across industries.

Neither sector is doing exactly the other's job and it often means a genuine partnership, one that has produced results neither could achieve alone.

Here we look at how that partnership actually works, and why both halves of it matter.

Artemis II

On 10 April 2026, four astronauts splashed down in the Pacific off the coast of San Diego. Reid Wiseman, Victor Glover, Christina Koch, and Jeremy Hansen had just returned from a ten-day journey around the Moon, the first crewed flight beyond low Earth orbit in more than fifty years. Their spacecraft, named Integrity, had flown 252,756 miles from home - farther than any human mission before it, beating the record set by Apollo 13 in 1970.

This was Artemis II - the first crewed mission in NASA's Artemis program, which aims to return humans to the Moon's surface. It was also one of the clearest examples of how public and private space now work together.

The rocket that launched them was NASA's Space Launch System. The spacecraft was NASA's Orion. But Orion's service module was built by the European Space Agency. One of the four crew members was Canadian, flown under an agreement with the Canadian Space Agency. And the lander that will carry Artemis III astronauts from lunar orbit to the surface - the next mission in the sequence - is being built not by NASA but by SpaceX, a modified version of the Starship rocket the company has been developing for its own Mars ambitions.

No single country or company could have pulled this off alone.

The mission also brought back new imagery of the Moon's far side and the first views of a solar eclipse seen from deep space, images no previous crew had captured. As with every crewed mission, there will be lessons about how astronauts live, work and communicate in deep space that will eventually find their way into our daily lives too.

Space exploration has always moved in cycles and throughout space history, even the most promising initiatives have met political headwinds and budget constraints.

The Apollo program reached its funding peak years before the first Moon landing, then saw resources decline even as astronauts were still walking on its surface. Space station plans, Mars mission proposals, and advanced telescope projects have all experienced the familiar rhythm of enthusiasm followed by fiscal reality.

What Private Companies Bring

For most of the space-age, the companies involved in space exploration were contractors. They built hardware to government specifications, got paid on a cost-plus basis, and went home. The rocket, the spacecraft, the mission belonged to the agency. That pattern has changed.

The companies in the headlines now are not just suppliers. They are designing their own rockets, running their own launch facilities, and setting their own priorities. Some of what they have produced is genuinely new, and it is worth looking at what they have actually done, because the story is more specific than "private space is rising."

The clearest example is SpaceX's Falcon 9. Before it, every rocket booster was thrown away after a single launch. Reusability had been studied for decades and most engineers assumed it wasn't practical. The Space Shuttle was partly reusable but never cheap. SpaceX proved it could be done at scale. Individual Falcon 9 boosters have now flown more than twenty times each, and launch costs have come down substantially as a result. That matters because cheaper launches mean more satellites, better weather data, better mapping, better communications, better science. NASA itself uses SpaceX rockets for many of its own missions, including the ones that carry its astronauts.

Rocket Lab has done something similar at a smaller scale. Before its Electron rocket began flying from a private site in Mahia, New Zealand, small satellites had to hitch rides on larger rockets carrying bigger payloads. That meant accepting whatever orbit the primary customer wanted and often waiting months or years for a launch slot. Electron offers dedicated rides to the orbits customers actually need. Universities, research groups, and smaller countries now have access to orbit in a way they didn't a decade ago.

Why can private companies drive costs down when decades of government programs couldn't? The reasons are structural rather than technical. A private company's survival depends on winning customers at competitive prices, which pulls every engineering decision toward cost reduction. A government program is accountable to taxpayers and elected representatives, which pulls decisions toward reliability, political durability, and fair distribution of work across contractors and constituencies. Neither set of pressures is wrong. They produce different things. SpaceX's willingness to accept visible failures during development, to consolidate suppliers rather than spread work across many companies, and to build reusability into the design from the start would have been politically impossible inside NASA. Just as NASA's willingness to spend decades on a single scientific question would have bankrupted a commercial operator.

These companies are not purely private in the way the headlines sometimes suggest. SpaceX has received tens of billions of dollars in NASA and Department of Defense contracts over the years. Commercial Crew alone was worth around $2.6 billion. NASA's Human Landing System contract with SpaceX for the Artemis lunar lander is worth roughly $4 billion. Public money is still doing a lot of the work. What has changed is how that money is spent, and what happens to the results. In the private sector this is often the Intellectual Property, or the IP.

What Public Agencies Bring

Public agencies produce things commercial pressure doesn't reach. Some of that is about choosing to do work that no customer is paying for because the knowledge matters on its own terms.

For example, the James Webb Space Telescope took roughly twenty-five years from concept to launch and cost around $10 billion. It was built by NASA in partnership with the European Space Agency and the Canadian Space Agency. No private company would have built it. The timescale is too long, the commercial return non-existent. Its purpose is to see deeper into the early universe than any previous instrument, and to share what it finds with the global research community.

Since Webb became operational in 2022, it has produced images of galaxies forming in the first few hundred million years after the Big Bang, the first direct detection of carbon dioxide in an exoplanet's atmosphere, and thousands of observations that researchers around the world will be publishing from for years to come. None of that has a paying customer. All of it changes what we know.

Landsat is the clearest single illustration of what public agencies produce that commercial pressure doesn't. Running since 1972, it provides the continuous, freely available baseline that commercial operators like Planet calibrate their own data against. Without that public record, the commercial one would still exist, but it would be harder to compare across time and harder to use for scientific analysis. The two approaches depend on each other.

This pattern runs through most of the examples in earlier chapters of this book. NASA's technology transfer program files around 1,600 new technology reports each year, signs somewhere between 140 and 180 new patent licences, and manages thousands of software usage agreements. When memory foam reached your mattress, scratch-resistant coating reached your glasses, or image-processing algorithms first developed for deep-space photography reached the mammogram machines in hospitals, it was because the program existed and was designed to push innovations into the wider economy. Private companies don't generally do this. Their IP is what they sell, or what keeps competitors out, and it isn't usually licensed cheaply to unrelated industries. What NASA runs is something different: a deliberate system for moving publicly funded knowledge into general use.

This isn't only a NASA story, although NASA is the biggest part of it. Other public space agencies around the world run their own versions of the same approach, and the picture of what public funding produces only makes full sense when you look beyond the United States.

Public Agencies: The Rest of The World

Most popular accounts of space exploration are written as though NASA is the whole story. It isn't, and it hasn't been for a long time. Public space agencies in Europe, Japan, India, China and a growing number of smaller countries have been doing their own ambitious work for decades, often in partnership with NASA and often alongside it.

The European Space Agency is probably the closest in scale and style. It's a cooperative of twenty-two member states, founded in 1975, and its record includes some missions that no other agency has matched. In 2014, the Rosetta spacecraft became the first to orbit a comet, and its lander Philae touched down on the surface — an achievement that had been in planning since the 1990s and required dozens of technical firsts. ESA's Gaia telescope has mapped nearly two billion stars in the Milky Way with a level of precision that has changed how we understand the galaxy's structure. ESA also runs Copernicus, a fleet of Earth-observation satellites whose data, like Landsat's, is free for anyone to use. Copernicus now provides much of the foundational imagery used by European agriculture, disaster response, and climate research.

Japan's space agency, JAXA, has built a particular specialism in something difficult: bringing pieces of other worlds back to Earth. The Hayabusa mission returned grains of rock from asteroid Itokawa in 2010, the first sample return from an asteroid in history. Hayabusa2 followed in 2020 with a larger sample from asteroid Ryugu, including material that has since been found to contain amino acids and other organic molecules. JAXA also operates the Kibo laboratory on the International Space Station, contributes to the Artemis program, and runs the GOSAT satellites, which measure greenhouse gases globally with enough precision to identify individual industrial emitters.

India's space agency, ISRO, has become a major space power in less than two decades. In August 2023, Chandrayaan-3 made India the first country to soft-land a spacecraft near the Moon's south pole, a region of particular scientific interest because of the water ice thought to be preserved in its permanently shadowed craters. India's Mars Orbiter Mission, Mangalyaan, reached Mars in 2014 at a cost of around $74 million, roughly a tenth of what NASA's comparable MAVEN mission cost the same year. ISRO also runs an extensive Earth-observation program that is particularly valuable for monitoring monsoon agriculture across South Asia, where a season's harvest can depend on accurate rainfall prediction weeks in advance.

China's space program has moved quickly. In 2019, Chang'e 4 made the first ever soft landing on the far side of the Moon, a technical achievement that requires communicating via a relay satellite because the Moon blocks direct signals. Chang'e 5 returned Moon samples in 2020, the first such return since the Soviet Luna 24 in 1976. Tianwen-1 put a rover on Mars in 2021. Since 2022, China has operated the Tiangong space station, which is permanently crewed and conducts a wide range of scientific research. Much of this work has been pursued without international partnership, partly by choice and partly because US law has since 2011 prohibited NASA from direct bilateral cooperation with Chinese agencies. The effect has been the emergence of a parallel space exploration tradition, running alongside the Western one rather than inside it.

The club has grown further still. The United Arab Emirates reached Mars orbit with the Hope probe in February 2021, making it the fifth entity ever to do so. South Korea launched its first domestically built rocket in 2022 and put its first lunar orbiter, Danuri, into operation the same year. Luxembourg has invested heavily in space resource law and commercial partnerships. Scotland's first orbital launch site is now under construction at SaxaVord in the Shetland Islands.

What this shows is that public space exploration has never really been a single-country effort. The benefits described throughout this book from the data, the scientific understanding, to the technology that eventually reaches daily life, come from a genuinely global enterprise. And the patterns are similar everywhere: long timescales, patient investment, and the deliberate choice to fund work whose returns are measured in decades or in knowledge rather than in next quarter's revenue.

Earth Observation

Nowhere is the public-private division of labour clearer than in Earth observation, because the two models are visibly running side by side, doing different jobs.

Public agencies provide the foundational data. Landsat, run by NASA and the US Geological Survey, has been imaging Earth continuously since 1972. Its satellites revisit each point on the planet roughly every sixteen days, at moderate resolution, and the entire archive is free. Europe's Copernicus program, run by ESA and the European Union, operates a fleet of Sentinel satellites doing similar work, also released openly through the Copernicus Open Access Hub. Japan's GOSAT monitors greenhouse gases globally. India's Cartosat series supports mapping and infrastructure planning across South Asia. Between them, public agencies produce a continuous global record of the planet that is available to anyone who wants to use it.

Private companies do something different. Planet, founded by former NASA engineers, operates more than 150 satellites each about the size of a shoebox, and together they image the entire surface of Earth every single day. Maxar's satellites produce high-resolution imagery of specific locations on demand, at a level of detail that public satellites don't match. These companies sell their data to governments, businesses, agricultural companies, defence agencies and investigative journalists. The business model depends on selling imagery, not giving it away.

There's one more piece to this that's easy to miss. Planet's satellites actually use Landsat and Copernicus as calibration references. Because public satellites have been running continuously for decades, with carefully maintained standards, they provide the benchmark that commercial operators measure their own data against. Without that public baseline, Planet's daily imagery would still exist, but it would be harder to compare across time, harder to stitch together with other data sources, and harder to rely on for scientific analysis. The public record isn't just running alongside the commercial one. The commercial one depends on it.

The two approaches aren't competing. They're doing different things. A climate scientist tracking how a glacier has retreated over thirty years needs Landsat's decades-long archive. A farmer who wants to know whether yesterday's storm damaged a specific field needs Planet's daily imagery. An emergency response team after an earthquake needs Maxar's high resolution. A forestry agency monitoring deforestation across an entire country needs Copernicus's open, consistent coverage. No single provider covers all of those needs, and no single business model would support all of them either.

What makes this work is that both halves exist. If public agencies stopped producing foundational open data, commercial operators wouldn't fill the gap - their business models don't support it. If commercial operators stopped producing high-resolution, high-frequency imagery, public agencies couldn't match it at that level of detail. The weather forecast you check, the climate projections that inform policy, the crop forecasts that shape food prices, and the satellite images you see on the news after a disaster all draw on both sources.

Commercial Crew

The clearest single example of the partnership working is a program that hardly anyone outside the space industry had heard of, until the day it put American astronauts back into orbit.

When the Space Shuttle was retired in 2011, NASA found itself in an unusual position. The agency that had put twelve astronauts on the Moon no longer had a way to get anyone to low Earth orbit. For the next nine years, every American astronaut travelling to the International Space Station did so on a Russian Soyuz rocket. The cost per seat rose steadily, from around $21 million in the program's early years to around $86 million by the last seats NASA bought. The arrangement worked, but it left the United States dependent on a foreign government for access to a space station the US had largely built and paid for.

NASA's response was to try something new. Rather than design and build a replacement spacecraft itself, which would have taken a decade and billions of dollars, NASA contracted private companies to do it. SpaceX and Boeing were each awarded contracts to develop crew vehicles that would meet NASA's safety requirements, using their own designs and their own approaches. The contracts were fixed-price, which meant that if the companies overran their budgets, they absorbed the cost rather than the taxpayer.

SpaceX delivered first. In May 2020, Crew Dragon carried NASA astronauts Bob Behnken and Doug Hurley to the ISS - the first crewed launch from American soil in nine years. The spacecraft has since flown regular rotation missions and has become the operational workhorse of American human spaceflight.

Boeing's Starliner took longer and ran into more trouble. After years of delays, an uncrewed test flight that had to be repeated, and technical problems that required extensive redesign, Starliner finally flew its first crewed mission in June 2024. That mission did not go to plan. Helium leaks and thruster failures on the flight to the ISS meant NASA decided it was too risky to return the crew in the same spacecraft. Astronauts Butch Wilmore and Suni Williams, who had launched on an eight-day mission, returned to Earth on a SpaceX Crew Dragon in February 2025 after nine months in orbit.

The Starliner difficulties are part of the story, not an exception to it. They show what happens when the commercial model meets a hard engineering problem that the schedule hasn't built in enough time to solve. But they don't invalidate the approach. Having two providers meant that when one ran into trouble, the other could step in. That redundancy is the point. And across the whole program, NASA has estimated that Commercial Crew cost the agency roughly a third to a half of what building an equivalent in-house system would have.

What Commercial Crew shows is that the partnership model can work, but only if both sides hold up their end. NASA set the destination and the safety bar. The companies solved the engineering problem in their own ways. Neither could have done it alone. It has meant that American astronauts are getting to the ISS again, at lower cost, on spacecraft NASA doesn't itself own, and the nine-year dependency on Soyuz is over.

Mars

If the partnership model has a test that no one can pass alone, it's Mars.

Robotic work has been going on there for decades. NASA's Perseverance rover has been operating on the Martian surface since February 2021, drilling rock cores and sealing them in sample tubes for eventual return to Earth. Its predecessor Curiosity, which landed in 2012, is still going. A small helicopter called Ingenuity flew 72 times before retiring in early 2024 - the first powered flight on another planet. ESA's Mars Express has been orbiting since 2003. Between them, Mars is the most thoroughly studied planet in the solar system apart from our own.

What hasn't yet happened is anything involving people. And the reason is that a crewed mission to Mars is an engineering problem on a different scale from anything attempted so far. The journey alone is six to nine months each way, with limited ability to turn around if something goes wrong. Astronauts would need radiation protection across that entire journey, in a way we have not yet built. They would need life support systems that can run for years without resupply, recycling nearly everything. They would need a way to land a spacecraft many times heavier than anything previously set down on Mars, through an atmosphere thin enough to make parachutes inadequate but thick enough to require heat shielding. And then they would need a way home.

No single organization has all the capabilities required. NASA has decades of planetary protection expertise, deep-space communication infrastructure, and experience of long-duration human missions on the ISS. ESA has contributed some of the most sophisticated orbital instruments ever flown. JAXA has particular expertise in sample return. SpaceX's Starship is designed around the cargo capacity that a Mars mission would need, though the vehicle is still in development and has not yet flown an operational orbital mission. Private manufacturing, government scientific expertise, international partnership, and long-term funding commitments: Mars needs all of them, from no one actor.

The clearest current example is Mars Sample Return. Planned jointly by NASA and ESA, the program aims to bring back the samples Perseverance has been caching since 2021. The scientific case is strong: returned samples would be analyzed in laboratories whose equipment is far more sensitive than anything that could be sent to

Mars, and might finally answer the question of whether the planet ever supported life. But the engineering is difficult, the cost has grown, and the timeline has slipped. In 2024, NASA opened the program to industry proposals for cheaper alternative architectures, signalling that even for a flagship science mission, the agency is now looking to commercial partners to bring costs down.

China's Tianwen-3 is pursuing its own sample return mission, currently aiming to launch around 2028 and return samples in the early 2030s. If it succeeds, it could bring Martian material back before the NASA-ESA effort does. That would be a first, and it would underline how genuinely global the Mars story has become.

When humans will actually reach Mars is a question no one can honestly answer. Working timelines have slipped repeatedly. NASA has spoken about crewed missions in the 2030s or 2040s. SpaceX has been more aggressive about timing. Neither should be treated as a settled plan.

What can be said is that when it happens, it will happen because several agencies, several countries, and several companies will have each contributed what they can. The idea that Mars might be reached by a single company acting alone, however well-funded, is unlikely to survive contact with the engineering reality.

The Ownership Question

Earlier in the chapter, we set aside a question. What has changed is how money is spent, and what happens to the results. This is the part of the chapter that answers it.

Under the older model, NASA not only paid the costs, it also owned what it found out. The rockets, the spacecraft, the designs, the manufacturing processes: all of it was a public asset. That meant the knowledge could be licensed broadly, shared with allied nations, or released into the public domain through the technology-transfer program described in the next chapter. When memory foam reached your mattress or scratch-resistant coating reached your glasses, it was partly because the underlying work was owned by a public agency with a mandate to share.

Under the newer model, NASA pays less and takes on less risk, but the contractor owns the result. SpaceX's reusable Falcon 9 is SpaceX's rocket, not NASA's. The specific engineering that made reusability work belongs to the company, not the public that helped fund its development. NASA retains certain rights such as government-purpose use, data access, and the ability to keep buying services at competitive prices, but the broader public benefit flows more narrowly than before. The innovations reach the market through the products SpaceX sells, not through a public catalogue that other industries can draw from.

There's also a legal backstop called march-in rights, written into US law under the Bayh-Dole Act, which allows the government to require that a federally funded invention be licensed more widely if the original owner isn't making it available on reasonable terms. In practice it has almost never been used, but it exists as a reminder that the public has some claim on what gets built with public money.

It's worth saying that the picture varies from contract to contract. Commercial Crew, the Human Landing System for Artemis, and the older cost-plus programs all have different IP arrangements, and some modern contracts are more generous to the public side than others. But the overall direction has clearly shifted. Neither approach is wrong. The older model produced public assets at high cost. The newer model produces private assets at lower cost. The question is what each model means for the pipeline of benefits that reaches the rest of us, the objects and technologies that we take for granted and that have filled every chapter of this book. Under the older model, that pipeline ran wide. Under the newer one, the flow is narrower, but it exists.

The partnership this chapter has described isn't a theory or a preference. It's how space exploration actually works now, whether that's an Artemis mission that couldn't have flown without ESA's service module and SpaceX's upcoming lander, or a farmer checking Planet's daily imagery against Landsat's long archive to see how a field is changing.

How much of the work is publicly owned, and how much privately, is not a technical question. It is a question about who the work is for. The next chapter looks at the pipelines that move knowledge from a space agency laboratory into a hospital, a kitchen, a pair of glasses, or a weather forecast.

10

HOW SPACE IDEAS REACH YOU

If you want to see how space ideas reach everyday life, it helps to start with something specific you already know. A mattress is a good example. NASA did not set out to improve sleep on Earth. Engineers were trying to protect astronauts and test pilots from impact forces. The material developed for that problem later moved into medical cushions, then consumer products, and eventually into mattresses and pillows sold around the world.

The same thing happened in less visible ways too. Landsat was not created so that farmers, insurers, or app developers could make faster decisions. It was built to observe the planet systematically from orbit. But once that long record of satellite data was made publicly available, it became the basis for all kinds of tools and services far beyond the original mission.

The same is true of handheld DNA-testing systems now used in hospitals, field biology, and infectious-disease response. The problem that helped drive that work was not a hospital problem at all. It was a space problem: how do you analyze biological samples in a place where there is no full laboratory, no spare bench space, and very little room for error? NASA engineers built tools for that environment, protected the underlying inventions, and then made them available for wider use. From there, companies adapted them into products for life on Earth.

You can trace the same broad journey through many of the examples in this book. What changes is not the basic story, but the route. Sometimes knowledge moves through patents. Sometimes through software. Sometimes through open public data. Sometimes through partnerships that begin before the technology is even finished. This chapter looks at those routes.

Public Benefit vs. Private Profit

When you check the weather on your phone or benefit from advanced medical imaging at your doctor's office, you are seeing the results of two different ways of handling innovation.

The difference between NASA and a private company is not just what they invent, but what happens next.

NASA, funded by taxpayers, has a public mission. Its longer-term research may not make money straight away, and in many cases that is not the point. It may be trying to answer a scientific question, solve a mission problem, or build capability that will matter years later. Because of that, NASA is also under pressure to ensure broader public benefit from what it develops.

That can mean more open sharing of data, more accessible licensing terms, especially for startups, and more effort to make useful ideas available outside the original mission. Landsat is again the clearest example. Once its archive was opened up, the value did not remain with the satellite program alone. Farmers could monitor crop health. Cities could plan growth more effectively. Disaster teams could map damage faster. Scientists could track environmental change across decades. Much of that value would never have appeared in the same way if the archive had stayed locked behind expensive paywalls.

The same principle applies to the publicly available weather data, which supports forecasting services people use every day. Soil-moisture data can help guide agricultural decisions. Climate records are used by scientists, planners, and public agencies tracking long-term change. NASA does not have to build every service that grows from this information. The wider value appears because other people can use it.

Private companies work differently. Their focus is usually on commercial return, competitive advantage, and shareholder value. They often develop things with a clearer market in mind, keep more of the detail to themselves, and charge for access to the result. That is how most business works. It can produce excellent innovation, but the benefits tend to travel through narrower channels.

Both models are important. Private firms often move faster, build more efficiently, and push hard toward practical use. Public agencies can support longer-term work, release more of what they know, and create value in places where the market would not have started. That difference runs through many of the technologies in this book.

Designed to Be Shared

In the private sector, innovation often ends in exclusivity. A company creates something new, patents it, guards it, and sells it as a distinctive product or service. Space agencies such as NASA, ESA, and JAXA have often worked differently.

Part of their job is not only to solve the original problem, but to make sure useful knowledge does not stop there. NASA's Technology Transfer Program is one of the main ways this happens.

One route is patent licensing. NASA offers several licensing options, including evaluation licenses that allow smaller businesses to test whether a technology is commercially useful before committing fully, as well as full commercial licenses for established firms. This gives a wider range of companies access to inventions that might otherwise remain inside a federal lab.

Some of the examples in this book came through that route. Memory foam moved from aerospace and medical uses into consumer products. The DeBakey blood pump moved into cardiovascular medicine. Environmental technologies such as emulsified zero-valent iron were licensed for groundwater remediation. In each case, NASA did not become the manufacturer. It developed, protected, and then made the technology available so somebody else could build on it.

Software is another major route. NASA creates huge amounts of code for mission control, modelling, simulation, navigation, data analysis, and design. Some of it stays highly specialized, but some becomes useful far beyond spaceflight. Through the NASA Software Catalog, software can be made available to businesses, researchers, universities, and government agencies, often at little or no cost.

Software can spread capability quickly. A tool developed to help with spacecraft design or mission analysis can later be used in aviation, engineering, mapping, environmental work, or product development without the long delay that often comes with manufacturing physical hardware.

Open data is another pillar. Beyond Landsat, there is weather data used by forecasting services, soil moisture data used in agriculture, and climate records used by researchers and planners. NASA does not have to turn each dataset into a consumer product for it to matter. Once the information is usable and available, other people can turn it into services, companies, and decisions that affect daily life.

Partnerships are important too. In some cases, technology transfer starts before a mission has even launched. NASA often works with companies during research and development, through programmes such as SBIR and through Space Act Agreements. That means firms can begin adapting ideas while they are still being developed, rather than waiting until the end. Astrobotic is one example on the lunar side. LED plant-growth systems later used in commercial settings are another reminder that some of these links begin very early.

The details of those partnerships are not always public in full. The existence of the work may be visible, while the technical and intellectual-property arrangements remain commercially sensitive. But the practical effect is clear enough. These collaborations can shorten the time between a space requirement and a useful Earth application.

From Space Lab to Market

One of the clearest ways to understand how this works is to follow a single case from beginning to end.

Engineers at NASA's Johnson Space Center faced a problem that sounds narrow until you think about what it means in practice. How do you analyze biological samples aboard the International Space Station without the luxury of a full laboratory infrastructure? On the ground, a lab might rely on pipettes, centrifuges, refrigeration, spare consumables, and trained technicians working in controlled conditions. In orbit, space is tight, crew time is limited, and procedures have to be safe, compact, and reliable.

NASA's answer was a patented, self-contained device for isolating DNA, RNA, proteins, and cells without depending on the usual array of bulky laboratory tools. It automated procedures that would otherwise be awkward and risky in microgravity.

Once developed and patented, NASA did not keep the idea locked inside its own programme. It was then made available through the Technology Transfer Program, opening the door to outside companies interested in healthcare, forensics, and field biology.

That did not produce a finished consumer product overnight. It rarely works that way. What it did do was make the core invention available so others could adapt it. Companies then refined the underlying idea for specific Earth-based uses.

One example is Cepheid's GeneXpert system, a rapid cartridge-based DNA analyser. During the COVID-19 pandemic, many people encountered such systems directly when receiving PCR tests that could produce results far faster than traditional laboratory pathways. The point is not that NASA invented every later diagnostic device. It did not. The point is that solving one hard biological-analysis problem in space helped generate technologies that others could reshape into products used in hospitals and wider public-health settings.

The same applies to related lab-on-a-chip and microfluidic systems first developed with Mars and ISS research in mind. In remote settings on Earth, where full laboratories are unavailable, those tools can be used for water testing, forensic screening, and food-safety work. In that sense, the route from space lab to market is not a single leap. It is a chain: mission problem, engineering solution, protection and release, outside adaptation, wider use.

Code That Powers Innovation

Patents are one way useful work travels. Software is another.

NASA creates a vast amount of code, much of it for highly technical purposes such as controlling spacecraft, modelling airflow, analysing mission data, simulating structures, and visualising complex systems. Some of that software stays inside aerospace. Some of it turns out to be useful much more broadly.

Cart3D is a good example and one we have already talked about. Developed at NASA Ames, it allows engineers to model how air flows around an object without building expensive physical prototypes at every stage of design. It was useful for spacecraft and aerospace work, but the principle is broader than that. If you can model airflow efficiently, you can use the same tool in aviation, motorsport, drone development, cycling, or any field where shape affects performance.

NASA chose to release Cart3D under an open-source license after it won Software of the Year. That meant the value did not stay confined to one agency. Engineers working in sectors far removed from rockets could use a tool first built for space-related work. The importance of that is not just that one software package found a second life. It is that software sharing can widen technical capability very quickly. A company or research group that could never afford to develop such a tool from scratch can still use it, adapt it, and build with it.

NASA's Software Catalog includes other tools that have found second lives outside spaceflight. WorldWind, for example, has been used as a visualization platform in government and research settings. Structural and simulation software first developed for aerospace work has also been used more widely in engineering.

Not every piece of NASA code becomes a consumer product. What it can also deliver is software written for difficult mission problems. This can later become usable by universities, companies, and other public bodies that could never have justified building the same tools from scratch.

Partnerships from the Start

Sometimes the useful link to Earth begins long before a technology is finished.

NASA often works with private companies during the early research and development phase, especially when a technology might be useful both in space and at home. These relationships can happen through Space Act Agreements, which give NASA a framework for working directly with outside partners, or through other support and development programmes.

That means the company is not simply waiting at the end to license a finished invention. It may already be involved in testing, adapting, and thinking about how the technology could be made usable beyond its first purpose.

Some of these relationships begin very early. Astrobotic, for instance, has developed lunar lander capability within NASA's wider contracting and support environment. In other cases, work that began with space-based research later found a wider commercial life. LED plant-growth systems developed for controlled growing in orbit are one example of a technology that later turned up in indoor and specialist growing environments on Earth.

Another example is NASA's work with Techshot and nScrypt on 3D bioprinting for the International Space Station. In orbit, the interest was in how tissues and related structures might be printed under microgravity conditions. On Earth, that same body of work may have implications for regenerative medicine and laboratory tissue engineering. The medical future of those systems is not settled, but the collaboration makes it much easier to explore than if the work had remained entirely within one public lab.

The point is not simply speed, though that's important too. It is that these partnerships bring manufacturing knowledge, commercial thinking, and practical adaptation into the story earlier. They can shorten the time between a difficult space requirement and a useful Earth application.

The Scale of Sharing

When you look at the numbers, this is not a side activity.

Every year, NASA inventors document around 1,600 potential innovations through what are called New Technology Reports. These are the first formal step in deciding whether something should be patented, catalogued, or developed further for wider use.

From these reports, NASA files dozens of patent applications each year. In 2024 alone, that included about 60 provisional applications and 80 non-provisional U.S. applications. The agency typically receives 60 to 70 new U.S. patents a year, adding to a substantial portfolio available for licensing.

It also actively puts those patents to work. Over the last few years, NASA has signed between 140 and 180 new patent license agreements annually. On the software side, the numbers are larger still. In 2024, NASA facilitated more than 5,500 software usage agreements.

That steady flow of reports, patents, licenses, and software releases shows that the process described in this chapter is systematic. These are not just lucky stories repeated because they sound good in publicity material. There is a structured effort to identify useful knowledge, protect it where needed, and make it available for wider use.

Licensing revenue feeds some of this back into the system. U.S. law and NASA policy set out how those royalties are handled. Inventors receive a share, and the NASA center where the work was done receives part of what remains. Those funds can be used to support further research, patenting, training, and technology-transfer work. It does not make NASA self-funding, and it does not remove the need for public investment, but it does mean that useful inventions can help support some of the wider cycle that produced them.

NASA's Economic Footprint

It is easy to look at the price tag for space missions and ask whether any of this is worth it. But unlike a private company, NASA does not exist to turn a profit. Its money comes through annual government appropriation.

In recent years, NASA's budget has typically been between $22 billion and $25 billion. For fiscal year 2023, it was $25.6 billion, which represented about 0.4 per cent of the total U.S. federal budget.

Looking only at the budget, though, misses much of the wider effect. NASA's own economic impact report for fiscal year 2023 estimated that the agency's activities generated more than $75.6 billion in total economic output across the United States, supported 304,803 jobs, and produced more than $9.5 billion in tax revenue across federal, state, and local governments.

Some of that came from large programmes such as Artemis and climate-related work, which drive contracts, manufacturing, and related services at scale. But the broader point is that NASA's spending does not stop with the mission itself. It moves through contractors, suppliers, research institutions, manufacturers, software firms, and all the secondary activity that follows.

That is part of why space is so difficult to judge simply by asking what one mission cost. Some of the return appears as science, some as infrastructure, some as capability, some as jobs, and some as products and services people later use in ordinary life.

Sharing Space Benefits Worldwide

This is not only an American story.

Other government-funded space agencies have built their own ways of making sure space research has wider social and economic uses. The structures differ, but the principle is recognisable. Work done for space can have value beyond space, and agencies have a role in helping that happen.

The European Space Agency runs a Technology Transfer Programme Office and a network of technology brokers that actively connect space-derived technologies with industrial and commercial uses across Europe. Japan's space agency, JAXA, also works with licensing, collaboration, and industrial use in ways that fit its national research and industrial goals.

Even within the United States, NASA's work is often multiplied by collaboration with other public bodies, including the Department of Defense, the Department of Energy, and the National Institutes of Health. Technologies do not travel in one simple line from one lab to one product. They move through a network of public research, private development, adaptation, and reuse.

That is one of the bigger achievements of the space age. A difficult, expensive, highly specialized piece of work is done for one reason, then made available in ways that allow other people to use it for another. That is how the benefits reach weather forecasts, hospital tests, engineering tools, protective materials, environmental systems, and products on a shop shelf long after the original mission is over.

11

DEBRIS AND DUST

On the morning of 5 November 2025, Commander Chen Dong was running final checks on the return capsule of Shenzhou-20, the Chinese spacecraft that had carried him and two crewmates to the Tiangong space station six months earlier. They were due to undock the next day and fly home.

Inspecting the viewport of the capsule, Chen noticed something that didn't belong. A small mark, triangular in shape, on the outer pane of the glass. His first thought, he said later, was that a leaf had somehow stuck to the window. Then he remembered he was in orbit, and that there are no leaves in orbit.

The crew used every tool on the station they could find to photograph the mark and sent the images down to Beijing. Within hours, ground teams had convened glass-materials specialists from across the Chinese space industry, run independent wind-tunnel simulations, and reached a conclusion: the mark was a penetrating crack, running from the outer surface of the window through to the inner pane. More than a centimetre long. The outer pane could detach during atmospheric re-entry, which would fail the pressure seal, which would depressurise the cabin and flood it with superheated gas.

The object that had caused the damage was estimated at less than one millimetre across. It had been travelling at something like seventeen thousand miles per hour.

Twelve hours later, the Chinese Manned Space Agency announced that Shenzhou-20 would not be flying its crew home. Chen Dong, Chen Zhongrui and Wang Jie would instead return aboard Shenzhou-21, the spacecraft their relief crew had ridden up. Three weeks after that, on 25 November, China launched Shenzhou-22 uncrewed, carrying food, supplies, and a specially designed patch to reinforce the damaged window on Shenzhou-20 so that the empty capsule could eventually be brought back to Earth on its own. It was the first emergency launch in the history of China's human spaceflight program. The damaged capsule finally parachuted down in Inner Mongolia in January 2026, two and a half months later than planned.

The crew were safe. The Tiangong station was safe. But the whole episode had been triggered by an object smaller than a grain of rice, which nobody had ever seen and which nobody could ever have tracked. Low Earth orbit, it turns out, is now full of such objects. That is the subject of this chapter.

The Scale of What's Up There

As of 2026 there are now about 140 million pieces of debris that size or smaller in low Earth orbit, according to the European Space Agency's 2025 Space Environment Report. Another 1.2 million are between one and ten centimetres across, large enough to destroy a satellite but too small to track from the ground. Roughly 54,000 objects are bigger than ten centimetres, of which around 43,500 are tracked continuously by the networks of ground-based radars and telescopes that keep tabs on everything large enough to monitor.

Then there are the working satellites. ESA's 2025 report supports roughly 40,000 tracked objects, about 11,000 active payloads, and a rapidly worsening debris picture.

The single biggest contributor to that increase is a system that did not exist a decade ago. As of early 2026, Starlink alone accounts for more than 9,400 operational satellites.

The US Federal Communications Commission has authorised the constellation to grow to 15,000, and SpaceX has applications pending for more than 22,000 further satellites beyond that, plus a proposed new 15,000-satellite layer at a much lower altitude of around 330 kilometres. Amazon's Project Kuiper constellation is adding to the total, Chinese operators are building their own constellations, and the European Union is developing a network called IRIS². Low Earth orbit is going from moderately populated to actively crowded, and the pace is still accelerating.

What this looks like in daily operations is relentless. Across the first six months of 2025, Starlink satellites performed 144,404 collision-avoidance manoeuvres. That is roughly one manoeuvre every two minutes, continuously, day and night, for six months. It was also three times the rate of the preceding six months. Each manoeuvre is the outcome of tracking an incoming close approach, calculating the risk, and firing thrusters to alter the satellite's path. This work is now so continuous that it has become a routine part of constellation operation rather than an unusual event.

Researchers at the University of British Columbia have tried to capture the trend in a single measure. Their Collision Realisation and Significant Harm clock, or CRASH clock, estimates how much time a satellite operator typically has to respond to a developing close pass. In January 2018, that window was 164 days. By June 2025, it had fallen to 5.5 days. The same analysis, covered by Scientific American, suggests that the average satellite in low Earth orbit now faces about a seventeen per cent chance of a close approach within any given 24 hours that could lead to a collision.

Kessler Syndrome

There's a name for what happens if the debris problem tips past a threshold. Kessler syndrome. It was proposed in 1978 by Donald Kessler, an astrophysicist at NASA's Johnson Space Center, and his colleague Burton Cour-Palais, in a paper with a deliberately technical title: *Collision frequency of artificial satellites: The creation of a debris belt.* The idea was simple. Once the density of objects in orbit crosses a certain point, collisions between them create more fragments, which in turn cause further collisions, which create more fragments still. The process, once started, can sustain itself without any further launches from the ground.

It has already happened on a small scale. In February 2009, an operational American communications satellite called Iridium 33 collided with a defunct Russian military satellite called Kosmos 2251 at an altitude of 789 kilometres. The two were travelling at a combined speed of around 11 kilometres per second. The impact destroyed both spacecraft instantly and produced more than 1,800 trackable fragments, many of them still in orbit today. That was a single unplanned collision involving two satellites. Imagine the equivalent mechanism working across dozens of objects in the same orbital shell, and the concern behind Kessler's paper starts to make sense.

What's disputed is how close we are to the threshold. Kessler himself wrote a follow-up paper in 2009, the year of the Iridium-Kosmos collision, concluding that his modelling now suggested parts of the low Earth orbit environment may already have crossed it. In other words: in some orbital shells, the debris population is probably generating new fragments from internal collisions faster than atmospheric drag can pull objects down. If that's right, the cascade has already started. It's just running slowly.

Other researchers are more cautious. Moriba Jah, an aerospace engineer at the University of Texas at Austin who studies orbital sustainability, has said publicly that he doesn't subscribe to the Kessler framing, not because he thinks cascading collisions can't happen, but because he thinks the framing oversimplifies a system that needs to be managed rather than merely feared. So the disagreement isn't about whether collisions will keep happening. They will. The disagreement is about how to think about a system that might already be past a tipping point in some places but not others, and that might be manageable in some orbital regimes but not in others.

There's one practical consequence of the cascade risk that's worth naming. In most engineering problems, a mistake is recoverable. You clean it up, redesign the process, try again. Orbital debris doesn't work that way. An object placed at 800 kilometres today can take centuries to come down naturally. The debris we have produced since 1957 is, for practical purposes, going to stay where it is on human timescales unless we go up and remove it. Which raises the question of who's going to do that.

Why The Market Alone Won't Solve This

Every company with satellites in orbit benefits from the satellites doing whatever they do: relaying signals, taking pictures, delivering data. But when one of those satellites fails at end of life and becomes debris, or when two objects collide and produce thousands of fragments, the cost isn't borne by the organisation that created the object. It's spread across every operator and every mission that now has to manoeuvre around the new debris. Economists call this a negative externality. Environmental writers call it a tragedy of the commons.

It's the same logic that makes air pollution difficult to regulate, or overfishing, or carbon emissions. Each individual actor has an incentive to take the private benefit of their activity. No individual actor has a direct incentive to bear the full shared cost. And the market, left to itself, doesn't produce a solution, because there's no customer to pay for the service of keeping orbit clean.

Consider what it currently costs to remove one piece of debris. ESA's ClearSpace-1 mission, due for launch in 2026, will attempt to capture a single defunct ESA satellite called PROBA-1 using a specially designed robotic spacecraft. PROBA-1 weighs 95 kilograms. The contract to remove it is worth 86 million euros. Even if future missions bring that cost down by an order of magnitude, removing the thousand most dangerous objects already in orbit would still require sustained investment in the billions of euros, with no commercial revenue to support it. Nobody pays to have debris removed from orbit, because nobody owns the orbit.

The situation is made more awkward by the fact that the actors most capable of cleaning up are also, in many cases, the actors most responsible for the growing traffic. SpaceX has built substantial engineering capability around satellite manoeuvre and deorbit. So have Iridium, OneWeb and Starlink's competitors. They could, in principle, be part of the response. But they're also the reason the response is needed. And their business models are built around launching and operating their own hardware, not capturing and removing someone else's.

There are other actors who could fund debris removal at scale. Governments could. National space agencies could. International consortia could. None of them currently do so at anywhere near the level the problem calls for. One reason is that the political case for spending billions on clearing space debris is harder to make than the case for spending billions on a new telescope or a Mars mission, because debris removal doesn't produce discoveries or pictures. It just prevents things from getting worse.

That is the bind. A problem caused by commercial activity that commercial activity alone won't fix. A problem that probably needs public funding for the same reason public funding produced Landsat and the James Webb telescope: because no customer exists who will pay for the benefit, even though the benefit is real and shared. The question the next section takes up is what's actually being tried, and how far it falls short of what the scale of the problem requires.

What's Being Tried

There are responses. None of them are enough, but they exist and they're worth describing honestly.

The most visible is ESA's Zero Debris Charter, launched in late 2023. The charter is a voluntary commitment by its signatories to design and operate spacecraft so that, by 2030, their activities produce no net addition to the debris population. This means reliable end-of-life deorbit, no deliberate fragmentation, and systems that can either dispose of themselves at mission end or be removed by someone else. As of early 2026, the charter has been signed by twelve countries and more than forty commercial operators and agencies, including Airbus, OHB, Thales Alenia Space, and a number of newer satellite operators.

What the charter isn't is binding. Signatories commit themselves publicly but face no legal consequence if they miss the target. It also doesn't address the debris already in orbit, only the addition of new material. And several of the largest operators in low Earth orbit are not signatories. SpaceX has not signed, though it has published its own debris-mitigation standards and generally operates its Starlink satellites in orbits low enough that atmospheric drag will bring failed units down within a few years without intervention. Chinese operators are largely outside the charter entirely.

Active debris removal is being attempted as well, though at a scale far below what the problem requires. ClearSpace-1, mentioned in the previous section, is one effort. Japan's Astroscale has successfully demonstrated several of the underlying technologies, including rendezvous with an uncooperative target and magnetic capture, through its ELSA-d mission. The UK Space Agency has funded feasibility studies for national debris-removal capability. A handful of other start-ups are working on similar approaches. All of these are useful. None of them, yet, are operating at a scale capable of keeping up with the rate at which new objects are arriving.

National guidelines are tightening. The old international rule of thumb was that satellites in low Earth orbit should be designed to deorbit within 25 years of mission end. In 2022, the US Federal Communications Commission shortened that to five years for new satellite authorisations within its jurisdiction, which is now a large share of global traffic. The European Space Agency is pushing for similar standards across its member states. Other jurisdictions have followed in varying forms. Five years is a meaningful improvement over 25. But for low-altitude constellations operating at 500 to 600 kilometres, where atmospheric drag naturally pulls satellites down in a few years anyway, the rule doesn't change much. The harder cases are higher orbits, where drag is negligible and objects will stay where they are for centuries unless they're actively removed.

A broader coordination effort, the Space Sustainability Rating, was launched in 2022 and now rates satellite operators on the sustainability of their practices. It's voluntary, like the charter, and its influence depends largely on whether customers and regulators decide to care about an operator's score. The early signs are cautiously positive but incremental.

Taken together, the current response is a mix of voluntary commitments, scattered technology demonstrations, tightening national rules, and rating schemes. It's not nothing. But it isn't matched to the scale of what's happening. The launch rate is rising, the collision-avoidance rate is rising, the CRASH clock is shrinking. The response is moving in the right direction, but more slowly than the problem itself.

The Regulatory Gap

There's a more fundamental question sitting behind all this, which is who governs orbit.

The framework currently in use was designed for a different era. The 1967 Outer Space Treaty, negotiated during the Cold War and still the foundation of international space law, was written when only two governments were putting anything into space. It says that states are responsible for the activities of their nationals, that weapons of mass destruction are not to be placed in orbit, and that space is the province of all humankind. It does not say what to do when a commercial operator in one country wants to put 15,000 satellites into orbits that will affect every other operator on the planet.

National regulators have stepped into the space that international law didn't anticipate. In the United States, authorisations for commercial satellite systems are issued primarily by the Federal Communications Commission, whose formal remit is managing radio spectrum. Because commercial satellites use radio frequencies, the FCC ends up making decisions about orbital placement, satellite numbers, deorbit requirements, and interference with other operators. This is not what the FCC was designed to do. It has become the de facto gatekeeper for commercial operations in low Earth orbit not by design, but because no other regulator had a clearer claim.

In January 2026, under chairman Brendan Carr, the FCC authorised SpaceX to expand its Starlink Gen2 constellation by a further 7,500 satellites, bringing the total authorised Starlink constellation to 15,000. The commission is now considering a further 22,488 Gen2 satellites, plus a separate proposal for 15,000 additional satellites at very low altitudes of around 330 kilometres. The chairman has described the approach as part of an "America first" strategy for competing with Chinese systems. Competing operators including Viasat, Globalstar, SES and O3b formally objected to the scale of the authorisation, citing orbital congestion and interference concerns. Those objections were largely overridden.

Two things are worth saying about this plainly. First, Starlink is a real service that millions of people now rely on, including remote communities, ships, aircraft, and aid organisations working in areas without ground infrastructure. The service has genuine public utility. Second, a single national regulator, operating under domestic political pressure and a remit that was never designed for global orbital governance, is now making decisions with consequences for every other country that uses low Earth orbit. The International Telecommunication Union, a United Nations agency with some authority over radio spectrum, has tried to push back on specific aspects of these authorisations, but its enforcement powers are limited and its processes are slow.

Other national regulators face their own versions of the same problem. China's programme, run through CNSA and the China Manned Space Agency, is subject to Chinese national law but operates outside the cooperative arrangements that govern most Western space activity. India, Japan, South Korea and the European Union each have their own regulatory arrangements, each reflecting their own national priorities. There is no single body with authority over orbit. There is no global permit system for placing hardware in low Earth orbit. The closest thing to international coordination is the UN Committee on the Peaceful Uses of Outer Space, which issues guidelines that nobody is obliged to follow.

The question isn't whether any of this is sinister. The question is whether a regulatory framework built in the 1960s for two governments, then adapted piecemeal through national agencies with their own domestic pressures, is the right framework for an orbital environment that now contains 10,000 operational spacecraft from dozens of actors and is adding thousands more each year. The honest answer is probably not.

The Questions This Raises

The purpose of this chapter isn't to predict what will go wrong. It's to show that space activity in the mid-2020s has developed in a direction that the partnership described earlier doesn't fully account for. There are real questions that the current arrangement hasn't settled.

Who pays when a satellite collision happens? Under the Outer Space Treaty, the country that launched an object is liable for damage it causes. But what happens when a satellite built by a company in one country, launched by a company in another, operating on a network run from a third, collides with something owned by a fourth? The liability framework wasn't designed for transnational commercial activity on this scale. To date, no major orbital collision has triggered a formal liability claim under the treaty. The first one that does will set a precedent, and nobody knows what that precedent will look like.

What happens when a major constellation stops operating? Every satellite in low Earth orbit has a finite operational life. Starlink satellites are designed to deorbit within around five years of end of service. The question is what happens to the 9,400 satellites currently in orbit, and the 15,000 authorised, if SpaceX itself were to go through a period of serious financial stress, or decide to abandon the service, or be forced to. The deorbit plan depends on active control of the satellites. If active control stops, the plan stops. None of this is a prediction, but it's a genuine systemic risk that no regulator has fully considered.

What about the orbits that don't clean themselves up? Low Earth orbit has the useful property that atmospheric drag pulls failed objects down within years or decades. Higher orbits do not. The geostationary belt at 36,000 kilometres, where most weather and communications satellites sit, has no meaningful drag, and objects placed there stay there indefinitely. The "graveyard orbit" used for retired geostationary satellites is filling up. There is currently no commercial or public programme to remove objects from either.

What does continuous commercial coverage do to the science? Astronomers have been raising concerns for several years that the sheer number of satellites crossing telescope fields of view is degrading observational astronomy. Starlink satellites leave bright trails across long-exposure images. Radio astronomy is affected by satellite communications in neighbouring frequency bands. The problem isn't abstract — professional observatories are now routinely reporting contaminated data, and the Vera Rubin Observatory, which began operations in 2025, is expected to have significant portions of its survey imagery affected by satellite streaks. This is a case where a commercial benefit that millions of people value — connectivity — is producing a cost that falls on a different community entirely, and the trade-off is being made without anyone having been asked to approve it.

And the question that runs underneath all of these: what happens to the public benefit this book has been tracing, when more of the space activity that produces it is captured within private systems?

We need to bear in mind that the commercial sector does real science and that private operators do conduct research on the ISS and on their own vehicles. Axiom Space has run more than thirty experiments per mission to the station, including investigations into cancer cell behaviour in microgravity, stem cell research, and studies of neurodegenerative disease. That work is important. But the findings belong to the companies that funded the research (or their partners), and not to a public catalogue that other researchers or industries can draw from. It means that the pipeline is narrower, as noted in the last section. And the narrower it gets, the more of space activity's benefits flow through commercial channels only.

None of these questions have clean answers. They are not the kind of problems that will be solved by a single piece of legislation or a single international agreement. They are the kinds of problems that will have to be worked out slowly, through negotiation, and through the gradual accumulation of working arrangements. The first step toward working them out is naming them.

What this chapter has tried to show is that the same expansion that produced those successes has also produced problems the partnership model doesn't yet know how to solve. None of these are reasons to stop. They are reasons to think carefully about how the next part gets done. And who it gets done for.

12

CONCLUSION

The memory foam in your mattress, the scratch-resistant coating on your glasses, the image-processing algorithms behind modern mammograms, the weather forecast you checked this morning, the satellite imagery that helps agricultural planning in your region and might have affected what you paid for bread last week: each of these traces back, in one way or another, to space research and the systems built to make its findings widely available.

None of this flow is automatic. It exists because choices were made, over decades, about how to organise space exploration and who should benefit from it. Public funding and private investment aimed at difficult problems with open data policies, technology transfer programmes, international cooperation and public-private partnerships. All of these were choices. And the questions raised in the previous few chapters, about debris, regulation, ownership and the distribution of benefit, are really questions about how those choices will continue to be made.

It isn't, in the end, a list of products. It is a way of producing and sharing knowledge.

The constraints of operating outside the atmosphere, the vacuum, the radiation, the temperature extremes, the absence of spare parts or easy repair, have forced the invention of techniques, materials and systems that then turned out to be useful back on Earth. Most of the innovations the book has covered reached us because the organisations that made them were structured to share what they found.

This combination of real problems and planned sharing is what created success. And it's not that space exploration 'happens' to generate useful byproducts - it's that a particular way of doing space exploration, with specific norms about openness, was designed to generate them, and it did.

The commercial sector has changed this pattern without entirely replacing it. SpaceX's reusable rockets have driven costs down to a level government programmes could not. Planet's daily imagery of Earth complements Landsat's long archive rather than competing with it. Rocket Lab has opened orbit to smaller players. These are real contributions.

But the commercial sector operates under different rules about what it shares. Its innovations reach the world through products its companies sell, not through a public catalogue others can draw from. Both models can work but they produce different kinds of flow, to different communities, at different speeds.

Where we are now

This is the landscape the book has been describing. In early 2025, the US administration proposed the deepest set of cuts to NASA in the agency's history which would have seen a reduction in science programmes, the termination of Mars Sample Return, the retirement of SLS and Orion after Artemis III, and the postponement of Lunar Gateway. The proposal was met with significant bipartisan pushback in Congress, and the final FY2026 settlement looked very different from the original request. SLS and Orion were protected, with congressional language explicitly preventing funds being moved away from them until commercial alternatives had proven they could match their capabilities. Most science programmes were preserved.

Some things did not survive. Mars Sample Return was formally cancelled, though $110 million was retained for what's now called Mars Future Missions. The sample tubes Perseverance has been caching since 2021 remain on Mars.

By the time this book is in your hands, the FY2027 budget proposal will have arrived. The cycle continues, which is what cycles do.

Some of those proposals will protect the pipeline the book has covered. Some will narrow it. Then again, we might discover the perfect balance of cost and service that opens up our future in space travel and in our understanding.

The Next Fifty Years

Nobody can predict how space exploration will develop between now and 2076. What we can do is look at what the choices are, and what direction the next fifty years could plausibly take.

For example, the pipeline might stay open. Public agencies keep doing long-term work on the problems commercial pressure prevent private companies tackling. Private operators keep driving costs down and widening access. The technology transfer mechanisms keep functioning. New innovations keep reaching hospitals, farms, weather services and living rooms. This is the continuation of the pattern we have seen across the book.

Or the pipeline might narrow as more activity moves into commercial hands and more innovations stay behind private IP walls. Public agencies retreat to the highest-risk scientific work, and much of what used to spread across industries through open licensing stays in the companies that produced it. The benefits keep arriving but through fewer channels, with more of the value captured by investors and customers rather than distributed broadly. This is what the ownership question was pointing at.

The environmental problems might tip the balance in unexpected directions. A major debris incident, a widespread cascade in a particular orbital shell, a collision involving a crewed spacecraft, could force regulatory and international arrangements that don't currently exist. Whether the response would be protective or restrictive is uncertain, but the current trajectory of launch rates, collision-avoidance manoeuvres and regulatory gaps can't continue indefinitely without some form of reckoning.

The partnership model might even mature. Contract structures might evolve that preserve more public benefit even when the contractor owns the 'knowledge' or IP. International coordination on debris and regulation might catch up with the scale of activity. New funding models might emerge that combine the patience of public investment with the efficiency of commercial operation. This is simply what happens if the current problems are taken seriously and addressed before they get worse.

No single one of these is inevitable, in fact all of them are possible. Which direction the next fifty years actually takes depends less on the technology and more on the decisions, mostly unglamorous, about how to organise and fund and regulate and share.

The ordinary objects in your life are the visible part of something larger. The cordless vacuum cleaner and the scratch-resistant glasses and the weather forecast and the GPS signal exist because a specific way of producing knowledge worked well enough, for long enough, for them to reach you.

You now know more than most people about where your ordinary things come from. What happens next is a matter for everyone who benefits from the flow, which means everyone reading this book. Whether the flow continues and in what form will be decided by the choices made by us all in the coming years.

But remember, the story of space exploration is, at its heart, a story about human ingenuity and drive. The scientists who built Webb, the engineers who landed Chandrayaan on the Moon, the astronauts who flew Artemis II, the people watching from their living rooms. None of them do this only for the spin-off technologies. They do it because looking outward is part of how we learn about ourselves. And above all else, for some reason that I think we can all understand, they love it.

BIBLIOGRAPHY

Abraham, W. T., Adamson, P. B., Bourge, R. C., Aaron, M. F., Costanzo, M. R., Stevenson, L. W., Strickland, W., Neelagaru, S., Raval, N., Krueger, S., Weiner, S., Shavelle, D., Jeffries, B., & Yadav, J. S. (2016). Wireless pulmonary artery haemodynamic monitoring in chronic heart failure: A randomised controlled trial. *The Lancet, 387*(10017), 419–427.

Axiom Space. (n.d.). *Axiom Space missions and commercial stations.* https://www.axiomspace.com/

Bauman, H. E. (1995). The origin of the HACCP system and subsequent evolution. *Food Science and Technology Today, 9*(2), 66–72.

Baylor College of Medicine. (2021, September 8). *Space Health Institute launches first commercial spaceflight medical research program.* https://www.bcm.edu/news/space-health-institute-launches-first-commercial-spaceflight-medical-research-program

Blue Origin. (n.d.). *Blue Origin official site.* https://www.blueorigin.com/

Canadian Space Agency. (n.d.). *NeuroArm.* https://www.asc-csa.gc.ca/eng/canadarm/neuroarm.asp

Center for Economics and Business Research. (2021). *What is the economic impact of the UK space industry?* UK Space Agency. https://assets.publishing.service.gov.uk/government/uploads/system/uploads/attachment_data/file/1020487/Economic_impact_of_the_UK_space_industry_-_summary_report.pdf

China National Space Administration. (n.d.). *Lunar exploration program.* http://www.cnsa.gov.cn/english/n6465652/index.html

Congressional Research Service. (2024, July 29). *Space resource extraction: Overview and issues for Congress* (CRS Report R48144). https://www.congress.gov/crs-product/R48144

Douglas, G. L., Zwart, S. R., Smith, S. M., & Rucker, R. B. (2021). Preparation strategies for long-duration space exploration: Recent improvements and current challenges. *Annual Review of Food Science and Technology, 12*, 551–577. https://doi.org/10.1146/annurev-food-062220-100836

European Space Agency. (n.d.). *Human and robotic exploration.* https://www.esa.int/Science_Exploration/Human_and_Robotic_Exploration

European Space Agency. (n.d.). *Technology transfer and patents.* https://www.esa.int/Applications/Technology_Transfer/Patents

European Space Agency. (n.d.). *Technology transfer programme.* https://www.esa.int/Applications/Telecommunications_Integrated_Applications/Technology_Transfer

Fossum, E. R. (1997). CMOS image sensors: Electronic camera-on-a-chip. *IEEE Transactions on Electron Devices, 44*(10), 1689–1698.

Frazier, O. H., Myers, T. J., Gregoric, I. D., Khan, T., Delgado, R., Spampinato, N., Radovancevic, B., & Cohn, W. E. (2001). First clinical use of the redesigned HeartMate II left ventricular assist system in the United States. *Texas Heart Institute Journal, 28*(3), 157–161.

Hargens, A. R., & Richardson, S. (2009). Cardiovascular adaptations, fluid shifts, and countermeasures related to space flight. *Respiratory Physiology & Neurobiology, 169*(Suppl. 1), S30–S33.

Huang, H., Roy, D. P., Boschetti, L., Zhang, H. K., Yan, L., Kumar, S. S., Gomez-Dans, J., & Li, J. (2022). The radiometric accuracy of the 8-band multi-spectral surface reflectance from the Planet SuperDove constellation. *International Journal of Applied Earth Observation and Geoinformation, 114*, 103015. https://doi.org/10.1016/j.jag.2022.103015

Hughson, R. L., Helm, A., & Durante, M. (2016). Heart in space: Effect of the extraterrestrial environment on the cardiovascular system. *Nature Reviews Cardiology, 15*, 167–180.

Indian Space Research Organisation. (n.d.). *Future missions*. https://www.isro.gov.in/

ISS National Laboratory. (2024, January 11). *Stem cell investigations to improve cancer treatment among payloads flying on Axiom Space's third private astronaut mission*. https://issnationallab.org/press-releases/release-ax3-life-science-research/

Japan Aerospace Exploration Agency. (n.d.). *Exploration*. https://global.jaxa.jp/projects/exploration/

Japan Aerospace Exploration Agency. (n.d.). *Martian Moons eXploration (MMX)*. https://www.mmx.jaxa.jp/en/

Jet Propulsion Laboratory. (2018, December 6). *Planetary defense: The Bennu experiment*. NASA Jet Propulsion Laboratory. https://www.jpl.nasa.gov/news/planetary-defense-the-bennu-experiment

Lauretta, D. S., Balram-Knutson, S. S., Beshore, E., Boynton, W. V., D'Aubigny, C. D., DellaGiustina, D. N., Enos, H. L., Golish, D. R., Hergenrother, C. W., Howell, E. S., et al. (2022). Spacecraft sample collection and subsurface excavation of asteroid (101955) Bennu. *Science, 377*(6603), 285–291. https://doi.org/10.1126/science.abm1018

Lunar and Planetary Institute. (n.d.). *Rocks tell us the Moon's story*. https://www.lpi.usra.edu/education/explore/marvelMoon/activities/familyNight/rocksSpeak/

Massa, G. D., Dufour, N. F., Smith, T. M., & Wheeler, R. M. (2017). Plant habitat: New hardware to support plant research for space crop production. *Gravitational and Space Research, 5*(2), 24–34.

Midwest Research Institute. (1971). *Economic impact of stimulated technological activity*. NASA.

NASA. (1972). *Space food technology* [Report]. NASA Technical Reports Server. https://ntrs.nasa.gov/api/citations/19720005422/downloads/19720005422.pdf

NASA. (n.d.). *Artemis program*. https://www.nasa.gov/specials/artemis/

NASA. (n.d.). *Automated isolation of nucleic acids, proteins, and cells*. NASA Technology Portfolio. https://technology.nasa.gov/patent/MSC-TOPS-42

NASA. (n.d.). *Habitable Worlds Observatory*. https://science.nasa.gov/astrophysics/programs/habitable-worlds-observatory/

NASA. (n.d.). *Kilopower project overview.* https://www.nasa.gov/directorates/stmd/tech-demo-missions-program/kilopower-hmqzw/

NASA. (n.d.). *Moon to Mars architecture.* https://www.nasa.gov/moontomarsarchitecture/

NASA. (n.d.). *NASA Technical Reports Server.* https://ntrs.nasa.gov/

NASA. (n.d.). *Spinoff.* https://spinoff.nasa.gov/

NASA. (2002). *Spinoff 2002.* https://spinoff.nasa.gov/

NASA. (2003). *Spinoff 2003.*

NASA. (2004). *Spinoff 2004.*

NASA. (2013). *Benefits stemming from space exploration.* https://www.nasa.gov/wp-content/uploads/2015/01/benefits-stemming-from-space-exploration-2013-tagged.pdf

NASA. (2015). *Food technology fact sheet* [Fact sheet]. https://www.nasa.gov/wp-content/uploads/2015/05/160398main_food_tech_fact_sheet.pdf

NASA. (2023). *Economic impact report: FY2023.* https://www.nasa.gov/sites/default/files/atoms/files/nasa-economic-impact-2023.pdf

NASA. (2023). *Spinoff 2023: NASA technology transfer success stories.* https://spinoff.nasa.gov/

NASA. (2023, June 28). *Long history and bright future of space sample deliveries. NASA Blogs: OSIRIS-REx.* https://blogs.nasa.gov/osiris-rex/2023/06/28/long-history-and-bright-future-of-space-sample-deliveries/

NASA. (2024). *Food safety program for space has taken over on Earth.* https://www.nasa.gov/directorates/stmd/tech-transfer/spinoffs/food-safety-program-for-space-has-taken-over-on-earth/

NASA. (2024). *Space food formulations: Thermostabilized* [PDF]. https://www.nasa.gov/wp-content/uploads/2024/06/nasa-space-food-formulations-thermostabilized.pdf

NASA. (2024). *Space food systems.* https://www.nasa.gov/directorates/esdmd/hhp/space-food-systems/

NASA. (2025, January 29). *NASA's asteroid Bennu sample reveals mix of life's ingredients.* https://www.nasa.gov/news-release/nasas-asteroid-bennu-sample-reveals-mix-of-lifes-ingredients/

NASA Science Mission Directorate. (n.d.). *Public access.* https://sti.nasa.gov/public-access/

NASA Software Catalog. (2024). *NASA software catalog.* https://software.nasa.gov

NASA Technology Transfer Program. (2021). *NASA technology helps detect and treat heart disease.* https://spinoff.nasa.gov/Spinoff2021/hm_1.html

National Air and Space Museum. (n.d.). *Lunar rocks.* https://airandspace.si.edu/explore/stories/lunar-rocks

National Oceanic and Atmospheric Administration. (n.d.). *Public value of NOAA's data.* https://www.noaa.gov/information-technology/public-value-of-noaas-data

Perchonok, M., & Bourland, C. (2002). NASA food systems: Past, present, and future. *Nutrition, 18*(10), 913–920. https://doi.org/10.1016/S0899-9007(02)00910-3

Planet Labs PBC. (n.d.). *Analysis-ready PlanetScope (ARPS) technical specification.* https://docs.planet.com/data/imagery/arps/techspec/

Planetary Society. (2025, March 5). *Breaking down Bennu: OSIRIS-REx finds life's building blocks in asteroid sample*. https://www.planetary.org/planetary-radio/2025-osiris-rex-sample

Planetary Society. (n.d.). *How do NASA and ESA work together?* https://www.planetary.org/articles/how-do-nasa-and-esa-work-together

Poulet, L., Fontaine, J.-P., & Dussap, C.-G. (2016). Plants' response to space environment: A comprehensive review including mechanistic modelling for future space gardeners. *Botany Letters, 163*(4), 337–347. https://doi.org/10.1080/23818107.2016.1194228

Quality Assurance & Food Safety. (n.d.). *It all started with NASA...* https://www.qualityassurancemag.com/news/haccp-origins-nasa/

Reuters. (2025, January 29). *Building blocks of life found in samples from asteroid Bennu*. https://www.reuters.com/science/building-blocks-life-found-samples-asteroid-bennu-2025-01-29/

Royal Society of Chemistry. (2017, March 7). It all began at NASA. *Education in Chemistry*. https://edu.rsc.org/feature/it-all-began-at-nasa-/3010375.article

Scientific American. (2024, February 20). *Medicine in space: What microgravity can tell us about human health*. https://www.scientificamerican.com/article/medicine-in-space-what-microgravity-can-tell-us-about-human-health/

Smith, S. M., Zwart, S. R., Block, G., Rice, B. L., & Davis-Street, J. E. (2020). Nutritional status assessment in spaceflight. *The Journal of Nutrition, 150*(9), 2233–2241. https://doi.org/10.1093/jn/nxaa086

Space Telescope Science Institute. (n.d.). *James Webb Space Telescope*. https://webbtelescope.org/

U.S. Department of Transportation. (2006). *Pavement grooving and skid resistance* [FHWA safety report].

U.S. Geological Survey. (n.d.). *Landsat missions*. https://www.usgs.gov/landsat-missions

Wall, M. (2025, March 11). Experts alarmed as White House proposes largest single-year cut to NASA in American history. *Space.com*. https://www.space.com/space-exploration/experts-alarmed-as-white-house-proposes-largest-single-year-cut-to-nasa-in-american-history

Wheeler, R. M. (2017). Agriculture for space: People and places paving the way. *Open Agriculture, 2*(1), 14–32. https://doi.org/10.1515/opag-2017-0002

World Intellectual Property Organization. (n.d.). *Intellectual property and space activities*. https://www.wipo.int/patent-law/en/pdf/developments/ip_space.pdf

Yiannas, F. (2009). *Food safety culture: Creating a behavior-based food safety management system*. Springer.

Zwart, S. R., Smith, S. M., & Heer, M. (2018). Nutritional biochemistry of space flight. *Advances in Clinical Chemistry, 87*, 179–252. https://doi.org/10.1016/bs.acc.2018.07.002

www.ingramcontent.com/pod-product-compliance
Lightning Source LLC
LaVergne TN
LVHW090413210926
848480LV00002B/550
9781068484841